SY 0086164 2

D1422375

POPULAR ENAMELLING

POPULAR ENAMELLING

Erika Speel

B.T. Batsford Ltd, London

Photographs by Stanley Speel
Line drawings by Janet Noad

First published 1984

ISBN 0 7134 4193 3 (cased)

Typeset by Servis Filmsetting Ltd, Manchester
and printed in Great Britain by
The Anchor Press Ltd
Tiptree, Essex
for the publishers
B.T. Batsford Ltd
4 Fitzhardinge Street
London W1H 0AH

Contents

Acknowledgements

I wish to thank the following friends, colleagues, organizations and Museums, for the assistance and facilities which were offered to me: C.F. *Barnes, Philip Barnes, Susan Graus, Hilde Hamann, Alan Mudd, Jane Short, Graus Antiques,* MCN *Antiques (Umezawa), the Ashmolean Museum, Oxford, The Bournemouth Museums, the Trustees of the British Museum, the Victoria and Albert Museum, and The Worshipful Company of Goldsmiths, London.*

The drawings were made by Janet Noad. The photographs (with the exceptions of those on pages 8, 87 and 92) were taken by Stanley Speel, and I am very grateful for having received this expert help.

ERIKA SPEEL

Introduction

The methods of enamelling, which are the techniques of fusing durable, coloured glazes to metal bases, have had an important place among the decorative arts for over 1000 years. The enamels can be applied and fired to the metal bases in many different ways and the designs and effects which can be created with the materials range from clearly defined patterns and smooth transparent or opaque glazes to narrative pictures and abstract compositions.

Within the scope of enamelling techniques it is possible to draw on the specialized skills of the decorative artist, the jeweller, the engraver, the metalworker or the miniaturist. Although most of the traditional enamelling methods were established over hundreds of years, enamelling is still a craft in which the materials can be adapted to personal requirements and an individual style can be created.

An engraving of a goldsmith-enameller's workshop of the Renaissance period. Courtesy of the Trustees of the British Museum.

1 Materials, Tools and Equipment

ENAMELS

Enamels which are suitable for fusing to a metal base are classified as Jewellery or Metal Enamels. With these enamels it is possible to give lustrous glazes with unfading colours to prepared metal bases.

Enamels are offered in many colours and in various qualities. They are broadly grouped under the headings of transparents or translucents, opalescents and opaques. The translucent enamels can be colourless or lightly tinted shades through which the surface of the underlying metal can be clearly seen or they can be more richly tinted and allow less light to pass through so that only very bright metal surfaces can show under the fired glaze. The opalescents are produced in a limited range of colours and with this type of enamel an irridescent sheen can be obtained in the fired glaze. The opaques give brilliant, dense glazes and they are produced in a very wide range of colours and shades.

A finished piece of enamelwork will have certain characteristics which are due to the method by which the work was made. Enamel can be recessed into the metal, in the inlaid techniques, or it can be fused to the surface of the metal in the encrusted techniques, or the colours can be applied in many thin layers in the painted techniques. The technique governs the range from which the enamels can be chosen, the way the enamels are prepared and applied and the number of stages needed to bring the work to completion.

Each enamel colour is unique, requiring a special formula and it may have subtle differences compared to other colours and qualities. For this reason it is helpful to look at some aspects of the manufacture of the enamels which are used by the craft enameller.

Manufacture of enamels

The manufacture of enamels is very specialized. The basic formula must result in a type of glass which will, when being fused or fired to a suitable metal surface during the enamelling process, form a stable bond with the metal. This means that the enamel must expand and contract on heating and cooling in a way which is compatible with the metal. The glassy base or fondant, which is the principal ingredient of all enamels, is known as flux and it looks like colourless glass. The flux is composed of silica to which salts of sodium, potassium, oxide of lead (lead-free qualities are also made) and borax are added in varying proportion. To make the coloured

Large lumps of enamel (frit).

Coarsely and finely ground enamel powder.

enamels, pigmenting agents are added to the flux during its manufacture. The colouring agents are metallic compounds, predominantly metallic oxides. Other minerals are added to improve the clarity of transparent colours or to impart varying degrees of opacity.

The raw materials for the making of each batch are melted together in a crucible or individual pot furnace until a glassy mass of the right elasticity is formed. The molten enamel is then poured from the crucible onto a metal slab to cool in small pools or cakes, or it can be discharged into a water bath so that it cools in small, rough lumps. Lump enamel is known as frit. The frit which has been produced by breaking up the cakes of enamel consists of smoothly surfaced lumps and in this state the enamel is non-porous and can be stored indefinitely without deterioration. Even in small rough lumps or in powder form enamel will not decompose for many years if kept dry. Enamel in frit or powdered form, particularly with transparents, will not show the true colour which can be obtained for enamelling.

Even very small changes in the proportions of the ingredients can alter the hue or shade of enamel. The range of colours which can be made is much greater than the number of colouring agents as different results are produced by combining various metallic oxides and altering the proportions of the ingredients. The metals which are used to colour enamels include: iron for green or red-brown; copper for various shades of turquoise, green and also for shades of red; cobalt for blues of various intensities; chromium for yellow-green; platinium for grey; iridium for black and grey; selenium for yellow or pink; antimony for yellow; nickel for brown; gold for transparent ruby red, pink, purple and opaque red; manganese for purple, tin for opaque white. A calx, or mixture of opacifying agents, is added to the raw materials to produce opaque enamels and the opaques consequently have slightly different properties than the transparents.

Some of the colouring agents tend to burn out at very high temperatures. Such pigments require a modified base (flux) formula which results in a lower melting point suitable for diffusing these colouring agents. These softer fusing enamels require more delicate firing for enamelling, in the range of 750–790°C (1382–1454°F), producing glazes which are hard and glossy in appearance but a little less resistant to acids and to scratching in subsequent use than the hard or high-fusing enamels. If too highly fired, the softer fusing enamels tend to loose good colour and will shrink, exposing parts of the metal surface. The higher fusing enamels are fired at temperatures above 800°C (1472°F) and they give the most brilliant and resistant surfaces.

Once made, the enamels cannot be mixed together to give intermediate colours. If powdered enamels are mixed they tend to fire with a mottled or speckled effect as the enamelling temperatures are too low to allow the pigments to blend. Colours can be modified or toned to some extent when enamelling by firing different layers over each other in successive stages.

Selecting a basic stock of enamels

It is generally necessary to have various qualities of certain of the colours particularly those to be used for the background or grounding.

Flux

A hard (high-fusing) flux will be needed for work on copper. Some qualities of flux tend to have a bluish tinge, others are slightly more amber and this will affect transparent colours fired over them. Silver flux is needed for a base of silver or silver foil. For overglazing a soft or super soft flux is required.

White

A hard white is needed for miniature painted work and this can also be used for larger scaled monochrome painted work. There is a special Dial White for clock and watch faces. Ivory or opalescent white are suitable for designs which require a warmer tone or a softer firing enamel.

Black

A small quantity of black is sufficient unless black-and-white monochrome designs are to be made.

Grey

Neutral shades ranging from light blue to true grey are offered in transparents and from light grey to dark grey and beige in the opaques. These are useful for offsetting more brilliant colours or to separate colours which would lack harmony if juxtaposed.

Blue

There are many beautiful blues in transparent and opaque qualities, giving some of the most outstanding of the enamel colours. Medium and dark Royal or Mazarine blue are particularly striking. The transparent blues generally improve with repeated firing.

Green

Transparent greens are offered in a good range, from light grass green to a dark bronze green. Emerald green is a very good transparent enamel. Some of the opaque greens are difficult to fire in thin, smooth layers, particularly on shaped articles.

Turquoise

Transparent turquoise may describe various tones from pale blue or green to a green-blue. These enamels fire extremely well over silver, giving very romantic colours. The opaque turquoise shades can be very fine. When underfired some of the transparent turquoise enamels appear opaque and overfiring the opaque turquoise enamels may produce some translucency.

Amber

The transparent ambers are very reliable enamels, they range from a light, golden colour to mid-brown. Most of the transparent ambers can be fired directly over a brightened copper base with good effect.

Brown

The darker, richly coloured transparent browns are the most effective.

Red

Some reds can withstand high firing, others turn muddy with repeated firing or the edges may darken. Over silver, transparent red should be fired over a grounding of silver flux. Cherry red in transparent and opaque qualities is very reliable.

Pink

Transparent and opaque pink require careful firing. Best results are produced if fired over a ground of opaque white.

Orange and yellow

These colours are produced in a range of shades in opaque qualities. Most are improved by firing over a grounding of flux or white.

A few of the transparents can be fired directly over a brightened copper base, but to produce a lighter and clearer transparent effect, a copper ground requires one or two fine layers of flux to be fired before the coloured transparent enamels are applied.

Counter-enamel

Any high-firing enamel or mixture of enamels can be used for counter-enamelling. Counter-enamel is a layer or layers of enamel fired to the reverse or counter-side of the metal. This stabilizes the work and counter-enamel is particularly important when a thin gauge of metal is used for the base or when the design is composed of a thick coating of enamel. If only

The metal tends to be pulled or warped out of shape when only one side is covered with enamel.

one side of a thin gauge metal base is covered with enamel the article is pulled out of shape, or warps, due to tension between the glazing and the metal. Counter-enamelling also reduces the risk of cracks developing in the glazing. With correct application and firing the counter-enamel remains fused to the metal despite the downward pull as the glazing re-melts at each firing.

Frit and powdered enamel

Enamel has to be applied in a finely powdered state. They can be bought in lumps (frit) or in a ready powdered form, in which case the frit will have been pulverized into a fine powder in a ball-mill. It is sometimes possible to specify when ordering from the manufacturers what fineness or grade of powder is preferred. The fineness is measured in mesh sizes – 80 mesh can be compared to cooking salt and 200 mesh is much finer, more like icing sugar. Because of the method of grinding, ready powdered enamels generally consist of a proportion of very fine particles mixed in with the stated mesh size. Transparent enamels tend to give more brilliant colours if not ground too finely. Opaques require to be very finely ground for good results.

Colours which will be required in larger quantities for regular use are

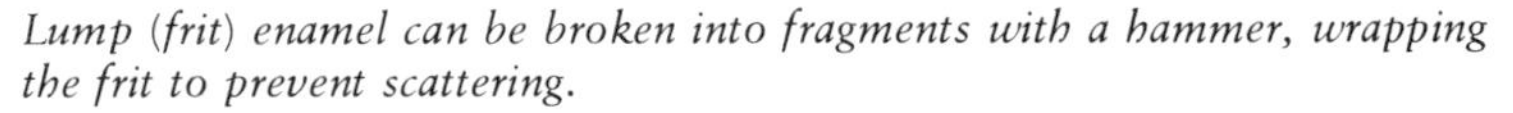

Lump (frit) enamel can be broken into fragments with a hammer, wrapping the frit to prevent scattering.

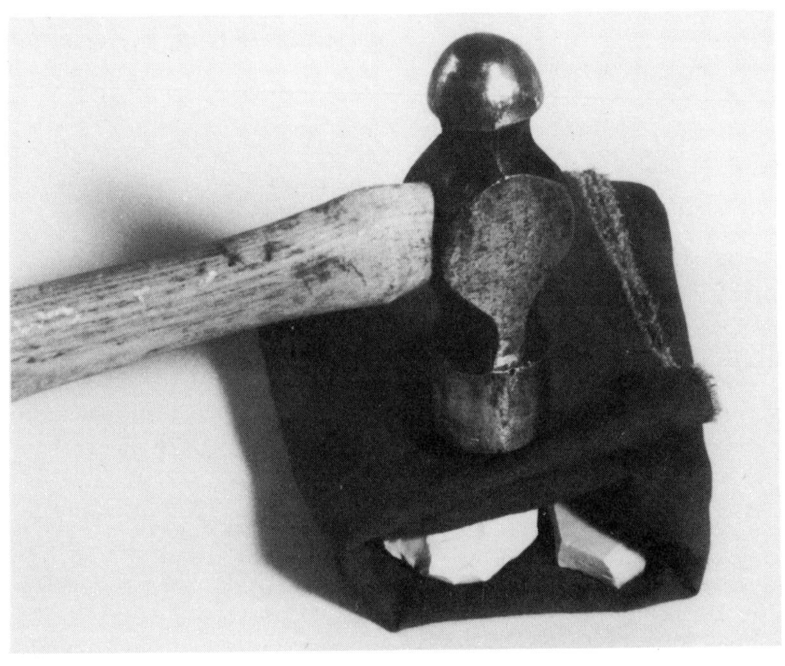

Grinding crushed frit into a powder with a mortar and pestle, keeping the frit well covered with water.

best bought in ready powdered form. Colours which will only be needed for small designs, for jewellery, and for use at intervals of time, can be bought in frit form and prepared into a powder of the necessary fineness prior to use. Hand grinding allows the powdered enamel to be prepared in the fineness to suit the job. Grinding and washing enamels immediately before use gives extra brilliance to the colours.

To grind frit into powder

Larger pieces of frit can be broken into fragments by wrapping them in strong paper or old cloth, to prevent chips flying off, and pounding with a hammer. When the frit is in pea-sized fragments, place a small quantity into a hard porcelain mortar and cover with water up to half the depth of the mortar. For larger quantities it is best to grind in two or three lots, otherwise it is difficult to produce evenly grained powder and it is best not to grind more than about 60 g (approx. 2 oz) at one time.

Place the mortar on a resilient surface or sand cushion to protect it from impact. The pestle is held in a palmar grasp with pressure applied to the top with the ball of the thumb. The enamel is pounded and crushed against the bottom of the mortar with the pestle. The frit gradually breaks into a sand-like texture. The pestle is then rocked and moved around the mortar to grind the enamel against the bottom and sides. As a finer consistency is produced, less pressure is needed and in the last stages the action is more of stirring. Throughout the grinding the enamel is kept well covered with water which, as it clouds with the release of precipitates, is poured off and renewed with fresh water. Before pouring off the cloudy water, tap the sides of the mortar gently with a wooden tool, settling the heavier enamel

During grinding the water under which the enamel is ground is poured off at intervals and renewed.

grains and pour off the water before the finer particles can sink again. For transparent and opalescent colours, washing several times until the water is quite clear during the various stages of grinding will improve the clarity of the glaze which can be produced. The grinding and washing reduces the amount of enamel which is left, but the quality of the remaining material will be improved.

After grinding and washing the powdered enamel, all excess water is poured off and the paste which remains in the mortar is transferred to a small dish with a spatula. Keep the paste covered while further colours are being prepared to prevent drying out and dust settling in. Note that if the

The prepared paste is covered when not in use to prevent drying out

paste does dry out slowly a crust tends to form on the surface and it is then best to re-grind gently and re-wash it before use. As long as the paste remains moist a few drops of water can be added as necessary to make up for evaporation. If the enamel is not required till a later date or is to be used in its dry state, it is best to dry it quickly on a piece of aluminium foil placed over a warm radiator or on top of the kiln and the perfectly dry powder can then be stored in an airtight jar.

Commercially pulverized enamels are usually supplied in a very fine powder and when this material is applied dry, by sieving or dusting it on through a fine mesh strainer (sifter), a very thin covering of enamel is deposited. This results in good transparency even when unwashed translucents are used so that the material can generally be applied directly as supplied for the dry methods. When pastes are required water has to be added to the powder and better results are produced if the translucents and opalescents are given several rinses with water as for the hand-ground enamels.

OVERGLAZES – METALLIC OXIDES

Metallic oxides can be bought in the form of fine powders. Quite a wide range of pigments is available. When applied to a glazed surface and fixed by refiring, the metallic oxides serve as dense overglazes or on-enamel paints.

Most of the metallic oxides can be mixed on the palette to produce intermediate shades, but some colours are not compatible. The colours can be made lighter by mixing in white. Colours can be made more fusible (requiring lesser heat to fix) by mixing with a special quality of very finely pulverized flux, known as printing flux. Printing flux added to a colour reduces the density and also gives these otherwise matt colours some degree of gloss.

To prepare the dry metallic oxide powders into an overglaze paint, each colour has to be mixed with a suitable painting medium. A colour is prepared by grinding a small quantity of the metallic oxide powder and the medium together on a glass slab (or glazed tile) using a flat-bottomed glass pestle (muller) or a spatula. The medium holds the pigment in suspension, to give a homogeneously coloured paint. This paint can be

Painting pigments require the addition of a medium and a small proportion of a thinning agent

Brushes for miniature work and application of lustres

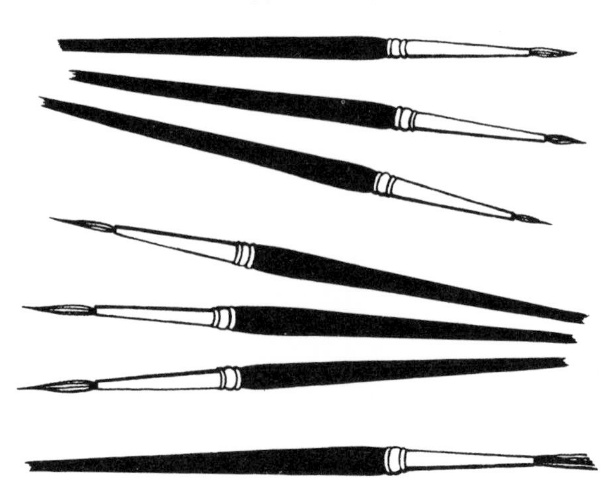

applied by brush to give finely controlled outlines, evenly spread colour or shaded effects. Spike oil of lavender, fat oil or a similar essential oil was traditionally used for the overglazes as they could dry off completely and did not affect the fired colour. A small proportion of a thinning agent, such as rectified spirits of turpentine could be worked into the paint to help it spread more easily. Modern proprietory painting media, such as screen-printing medium, can be substituted and these will dry off at a faster rate than the thicker oils, which shortens the time required to complete the painted work. A range of overglaze pigments is offered, with the colours in tablet form, which require only water for mixing and they are easy to apply, giving pale, gentle hues which fire with some gloss.

Metallic oxides are most suitable for painting of small-scale and miniature work. The materials are not washed so there is no waste and only sufficient colour is made up for the job in hand. As the materials are used very sparingly, metallic oxides are bought in very small quantities.

A basic stock of metallic oxides for on-enamel painting should consist of: black, white, yellow, maroon, scarlet, green, dark blue and rich brown, plus printing flux.

METALS FOR ENAMELLING

The craft enameller works with gold, silver, copper and some of their alloys, and in more limited ways with bases of steel. For commercial production a wider range of metals can be brought into use and as well as gold, silver, copper, copper alloys and steel, special qualities of enamel can be used on bases of iron and aluminium.

Gold

The melting point of pure 24-carat gold is 1063°C (1945.4°F). Gold is soft and malleable to work with and it does not oxidize. For enamelling purposes gold can withstand repeated firing. Due to the high cost and its softness, the use of pure gold is confined to use on small-scale jewellery pieces and the refurbishing of antique articles. For more general purposes 18-carat gold is used, which is less soft, less yellow than fine gold and has a melting point of 905°C (1661°F).

Silver

In Britain and the US the term fine silver may only be applied to metal of at least 999 parts in 1000 pure silver. Fine silver, being very reflective, gives the most brilliant effects with transparent enamels. It melts at 961.5°C (1762.7°F). Sterling silver (925 parts in 1000 of pure silver, containing copper as an alloy) melts at 893°C (1639.4°F). Sterling silver does not withstand multiple firing as well as fine silver.

A base of silver should be prepared by engraving or roughening to allow the enamel glaze to key into it and improve adhesion. During firing some enamel colours may react with a silver base and require firing over a special quality of 'silver' flux.

Enamel on a base of gold. Courtesy of the Victoria and Albert Museum.

Translucent enamel over silver.

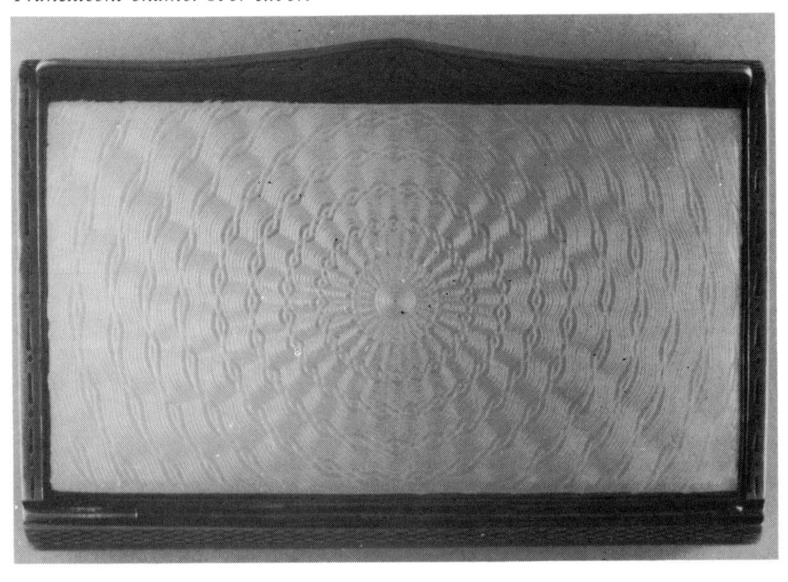

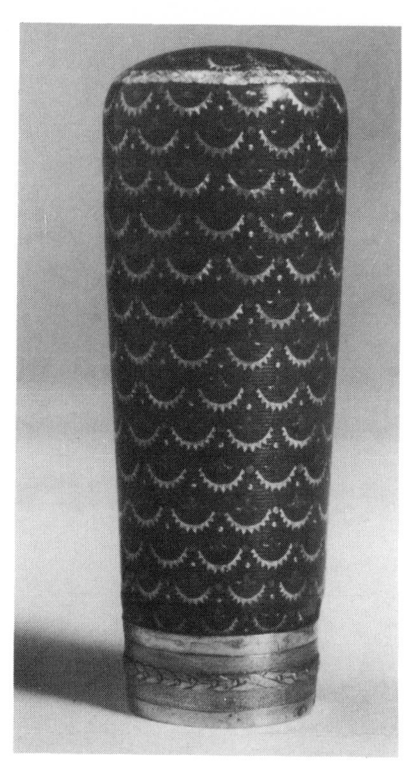

Translucent enamel on a base of silver gilt with embedded gold foil spangles on a lady's walking-cane handle.

Champlevé enamel on a base of copper.

Gold and silver foils

Gold and silver can be bought from specialist suppliers in the form of very thin foils suitable for enamelling. Foils are used under transparent or opalescent enamels to increase translucent effects or to alter the tone of a colour. Shaped foils or spangles can also be applied to the surface.

Copper

When clean and bright, copper is pinky-brown in colour. It has a melting point of 1083°C (1981.4°F). Copper is suitable for all the enamelling techniques and can withstand repeated firing. The parts of the copper surface which are not protected by a covering of enamel become oxidized as a result of high firing, and a dark, scaly layer forms on the metal surface, which has to be removed after each firing. To increase translucent effects, a copper surface is cleaned, brightened and given a grounding of flux before translucent enamels are applied.

Gilding metal

Gilding metal is reddish-golden in colour. It is an alloy of copper and zinc and for enamelling purposes the zinc content should be no greater than

Enamel on copper alloy base.

10% and preferably only 5%. Gilding metal of enamelling quality has a melting point of 1065°C (1950°F). It is most suitable for work which requires three firings or less. A few of the transparent colours, notably red and green, give particularly good effects when fired directly over gilding metal. As its name implies, the metal can be gilded after the enamelling is complete.

Steel

For enamelling a special form of mild steel is required. The melting point is

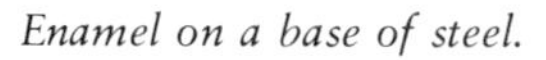

Enamel on a base of steel.

about 1350°C (2430°F). With steel only surface enamelling is possible as the metal cannot be worked by the usual studio methods. Steel requires special preparation to ensure a good bond with the enamel. As it is a non-reflective metal, opaque colours give the best results. A range of colours made specifically for use over steel is available.

Wires

For cloisonné, filigree and larger scale designs with metal outlines, flat, round, beaded or twisted wires are attached to the metal base to form cells or to zone areas of colour.

Gold, silver, copper or brass wires are suitable for these enamelling techniques. The wires can be of a higher quality than the metal used for the base of the work, with gold wires attached to a silver base and so on. The gauge or thickness of the outlining wires should be in proportion to the delicacy of the design and thickness of the base. For most enamelling purposes the wires are chosen from the range of Gauge 18 on the Standard Wire Gauge, approximately 1.22 mm (0.048 in) diameter to Gauge 24, approximately 0.56mm (0.022in) diameter. The noble metals can be used as extremely fine ribbons cut from gold or silver foil for very delicate work. Wires can be bought as round, flat or square in section. Round wire can be

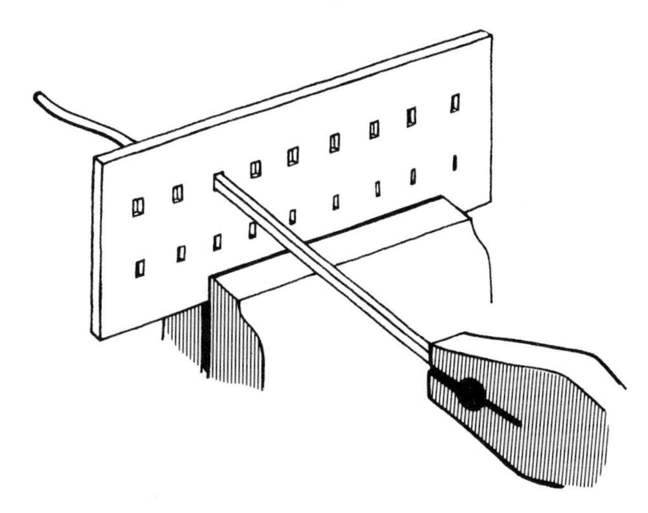

To change the gauge or shape of annealed wire it is pulled through successive holes in a drawplate

reduced in diameter or its section can be changed by pulling the annealed wire through a suitably shaped drawplate (a template of steel with a series of graduated holes of the desired shape, the wire being worked through decreasingly smaller or flatter holes). Round wire can also be flattened by hammering the annealed wire on an anvil.

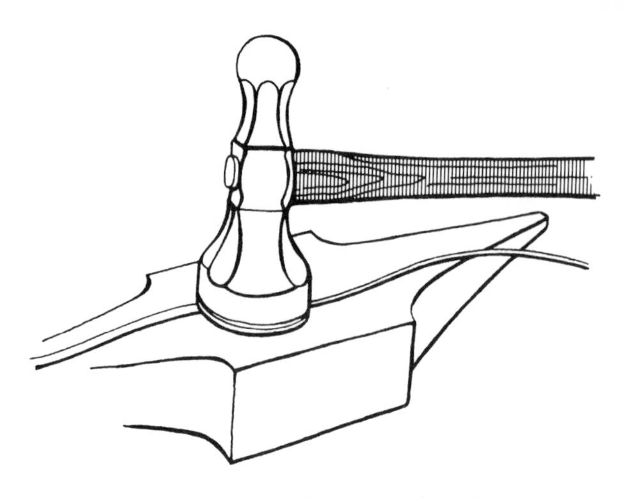

Annealed wire can be flattened with a hammer and anvil

THE ENAMELLING KILN

An enamelling kiln must be able to reach temperatures of 700–900°C (1292–1652°F) or higher if larger work is to be produced. The kiln has to be well insulated so that it is possible to work close up to it and to prevent heat loss which would make its operation slow and very expensive. The

a *Pyrometer – a calibrated indicator connected with a thermocouple in the muffle which can regulate the temperature*
b *An enamelling kiln with a vertically opening door which has a large inspection hole*
c *A metal shelf which is placed in front of the kiln for resting enamelwork before firing*

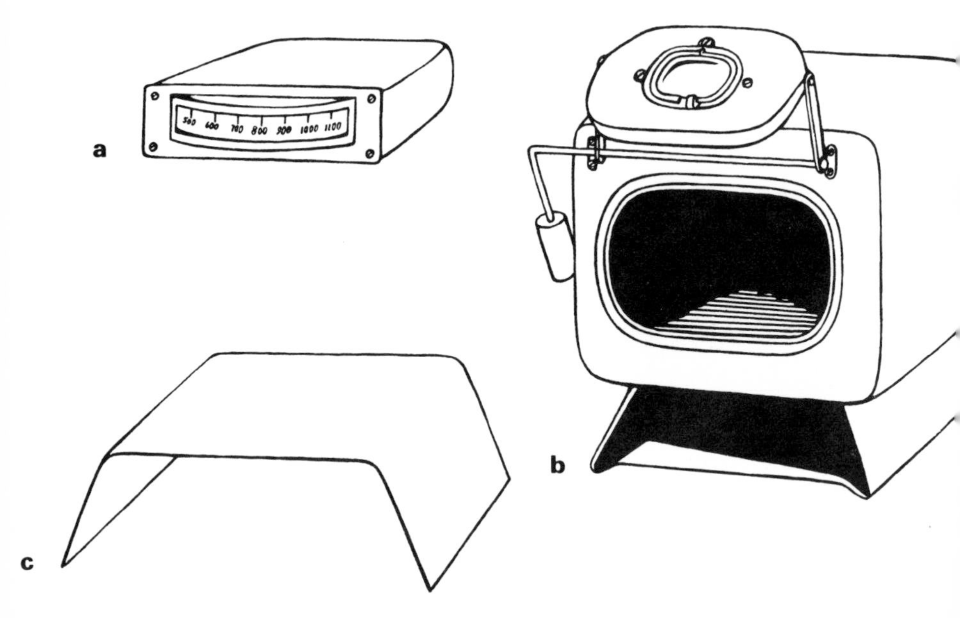

inner chamber, or muffle, may be considerably smaller than the external dimensions because of the thickness of the insulating walls.

In the past when other types of furnaces were in use, fired by charcoal, coal, gas or coke, enamelwork was placed into a removable clay muffle which was put into the firing chamber of a furnace. The muffle protected the work from the direct flames of the burning fuel and from smoke and soot while at the same time ensuring even, all-round heating. Modern electrically heated kilns have fixed interiors, or muffles performing the same function.

Small pieces of enamelwork can be successfully fired by means of heating torches or blow torches. When firing by such means the flame should only be directed at the underside or the base of the work and a piece of steel can be arched over the work to spread the heat more evenly. Counter-enamelling is difficult when firing by these means.

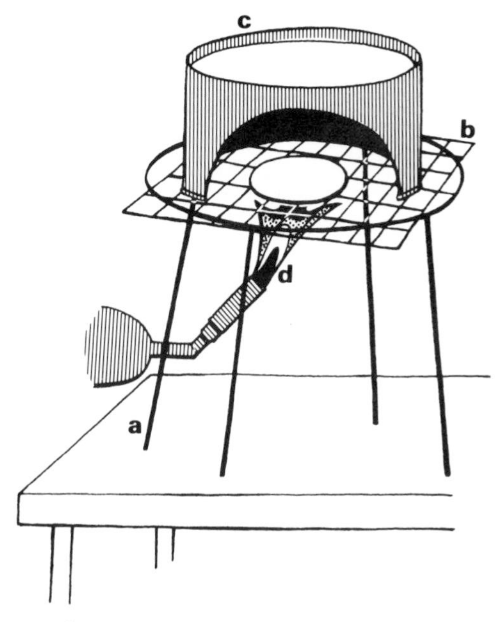

Firing by means of a torch:
a *Laboratory stand*
b *The plaque is supported on a mesh square*
c *Metal arch or hood to diffuse heat*
d *Flame applied from below*

Electrically heated kilns should have separate fuse boxes. The heating elements should be completely encased in non-conductive material, otherwise the kiln must have an automatic cut-off system which switches off the power when the kiln door is open.

When firing there should be good clearance within the muffle to ensure even heating.

While being fired, enamelwork must not be in contact with the sides or roof of the muffle and should be well clear of the floor and the door, fitting easily into the space available. Space is required around work being fired to accommodate the firing support and to allow for some expansion of the metal and for circulation of hot air. These factors should be considered when choosing a kiln.

The kiln has to be front-opening, with a door which is counter-balanced so that it can remain open without being held. The designs vary: doors may open vertically upwards or horizontally to either side, some opening forwards towards the operator. Ease and convenience of opening and shutting of the kiln door is very important as enamelwork is inserted into the hot kiln and is removed shortly afterwards when timing is critical, and each piece of enamelwork is fired several times. The kiln requires its own metal support or stand to hold it above the working table-top. There should be a lip or separate metal shelf at the entrance on which work can rest before firing. An inspection hole in the door is useful. A temperature gauge is necessary for work on gold and silver and it is useful for all classes of work. There should also be a means of holding the temperature at a set maximum heat, such as a thermostat or special switch to safeguard against overloading the equipment.

Finally, it is recommended that the cost of replacement parts should be considered as these can vary, some kilns being expensive to refurbish.

Firing tools

Firing tools, which are used to insert work into the hot muffle and to withdraw the glowing hot pieces, should have long handles with wooden or insulated grips. Tongs are more suitable for heavy pieces and for certain shapes of firing support where it is difficult to slide a spatula under, but they are more difficult to manipulate. Long, flat-bladed spatulas or two-pronged forks are generally used and these enable work to be set down gently without jarring.

Firing is safer and more comfortable if a heatproof glove is worn when placing work into the kiln. The soft, heatproof gloves with cuffs, made for boiler maintenance, are very suitable. An old, lined leather glove can substitute for a firing glove.

Firing supports

Supports hold work in the correct position during firing and raise the work off the floor of the muffle to ensure good circulation of heat. When both sides of the work are covered with enamel, the support has to be a stilt or cradle, the enamelwork resting on the points of the stilt or held by its rims in a U-shaped cradle.

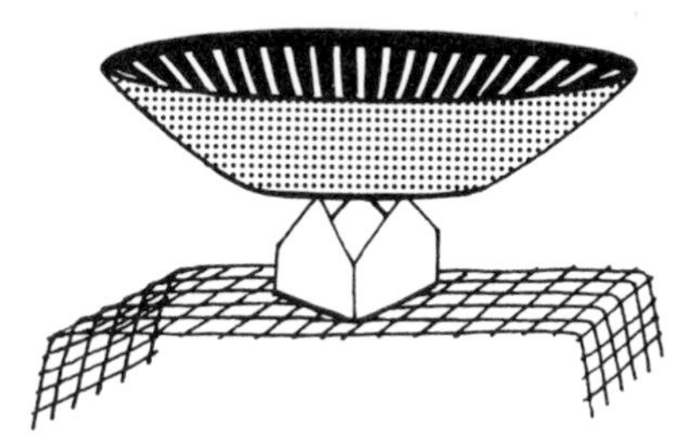

Stilt support

Supports can be made from various materials which can withstand repeated firing at enamelling temperatures. A variety of supports is offered by craft suppliers. To make up individual supports, perforated sheet steel or iron screening are suitable. Small supports, for buttons or ear-rings, can be made from fire-cement. Fire-cement is obtainable from builders' merchants in small quantities. This cement can be moulded into small brickettes, hollowing the base, with three steel pins inserted in the top, point upwards, before the material hardens. The fire-cement supports

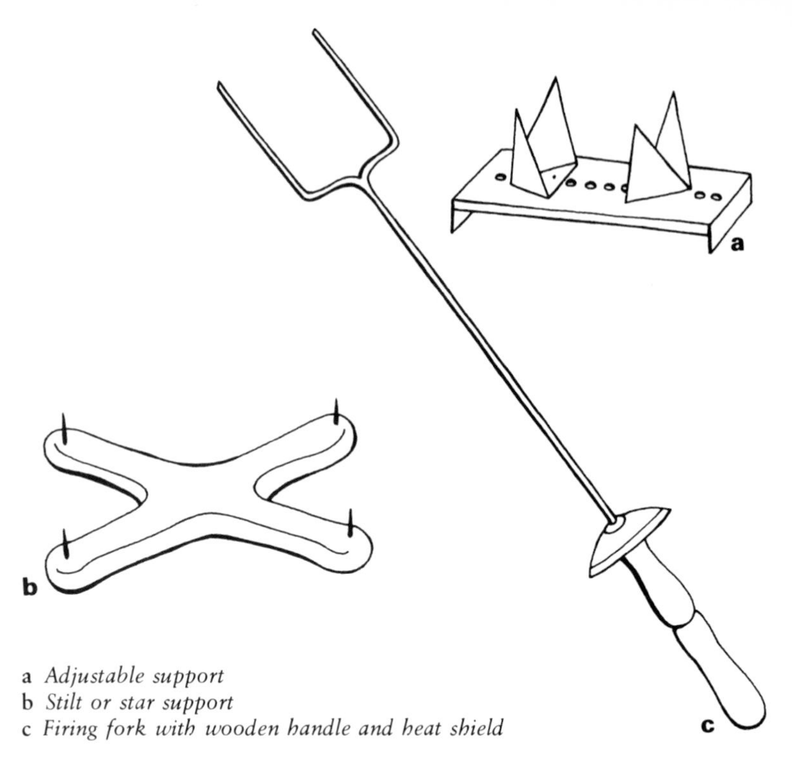

a *Adjustable support*
b *Stilt or star support*
c *Firing fork with wooden handle and heat shield*

have to dry out for 24 hours and should be prefired before use with enamels. For finger-rings or pieces which cannot stand easily in the required position, a piece of thick iron wire can be bent into a hook or sloping arm to suspend or support the work during firing. Note that when supports are made of iron the surface should be painted with a paste made from powdered jeweller's rouge and water, then dried and prefired, to prevent formation of fire-scale which could contaminate the enamelwork.

Any support which is small or difficult to lift up with the spatula can be placed on a flat tray support for easier handling.

Cooling slab

Enamelwork is removed from the kiln in a red hot state and needs a suitable cooling-off surface. A very thick steel plate or slab is most suitable. Thinner sheet metal can serve if placed over bricks. The cooling slab should be near to the kiln so that it is not excessively cold when the work is placed on it. The cooling area has to be kept clean and preferably should be on the opposite side to where work is waiting to be fired.

Weight

When only one side of flatwork is covered with enamel and when large plaques are being made, an old-fashioned flat iron or a defunct electric iron can be used to weight the work down during cooling to prevent warping.

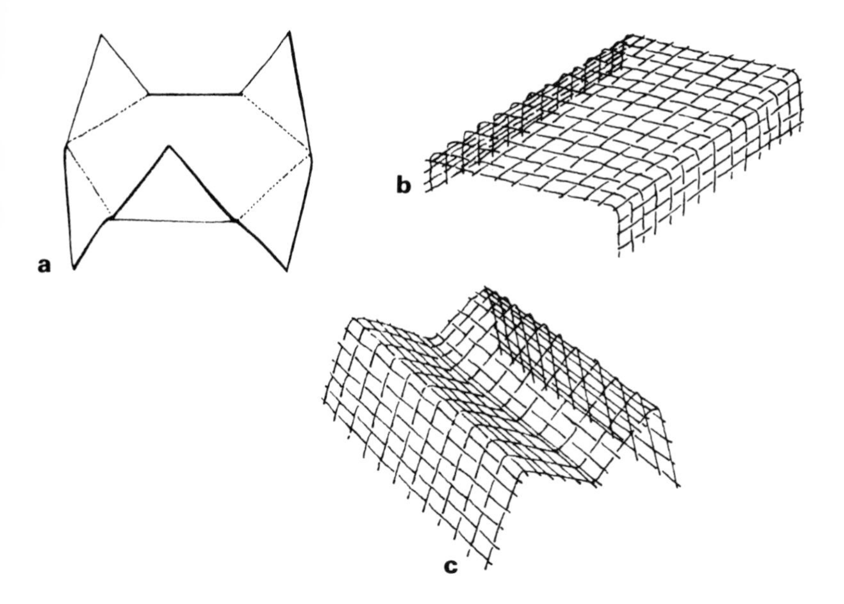

a *Stainless steel stilt support*
b *Flat tray mesh support – can be used in an inverted position to make a U-shaped support*
c *W-shaped mesh support – when used inverted the arms can be angled to fit various shapes*

Suitably shaped work, such as a ring, can be suspended for firing.

TOOLS AND AIDS FOR APPLYING ENAMELS AND OVERGLAZES

To apply wet enamel pastes

A sharpened quill, made from a goose or swan wing feather, is a traditional enamelling tool. The thick end of the feather's shaft is cut at an angle of about 45° with a sharp penknife or scalpel. This leaves a hollow cylinder with an angled, slightly flexible point which is very suitable for lifting, depositing and spreading the enamel paste into the design. Another traditional tool which can be made in the workshop consists of a small copper spatula made by taking a short length of thick copper wire and hammering one end into a flattened point and the other end is filed to leave a sharp point. However, many people prefer to use dental probes, pointed bamboo sticks, fine spatulas or brushes with which to work the pastes into the designs.

Some of the tools used for enamelling, including a spray (atomizer), pestle and mortar, paint brushes, fine-mesh sieve, tweezers, spatula, metal probe (point) and quill, copper wire, jeweller's pliers, wire cutters, glass brush, graver, carborundum stick.

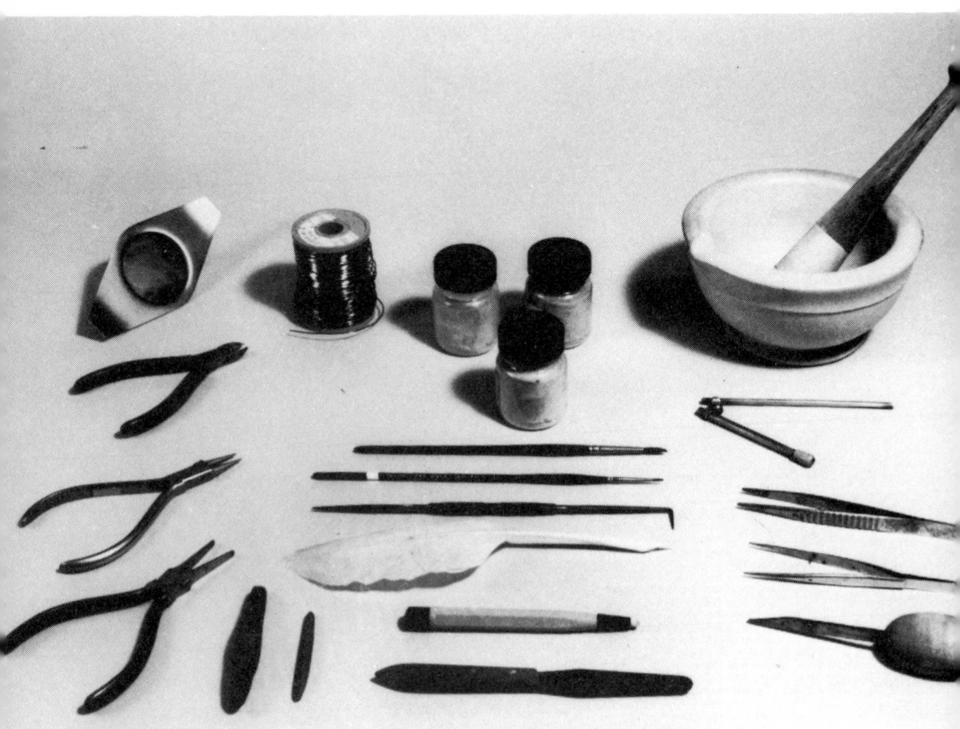

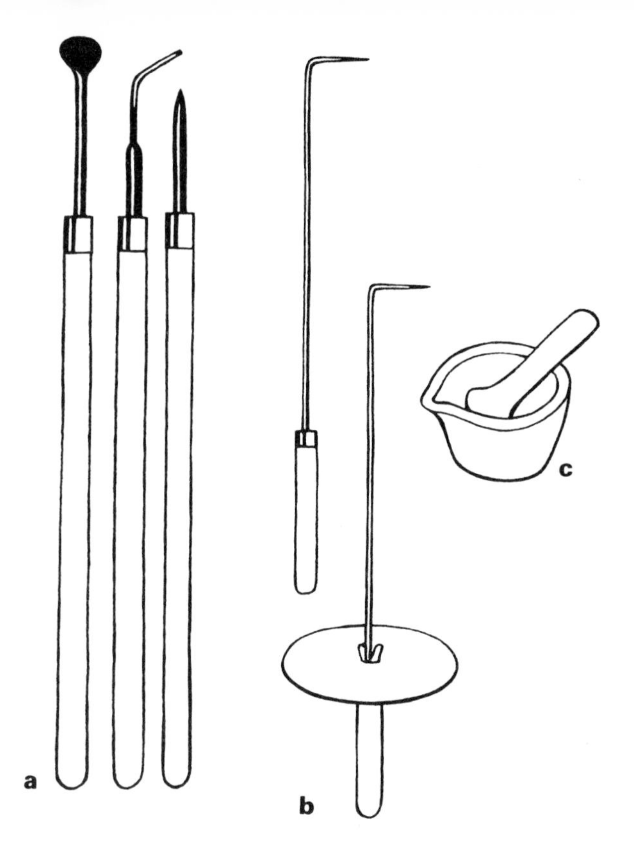

a *Inlaying tools*
b *Scrolling tools*
c *Pestle and mortar (not to scale)*

To apply dry enamel powder
Dry enamel has to be dusted over the metal surface as evenly as possible and this is quite easy to do by passing the powder through a sieve. A piece of fine copper or steel screening mesh can be shaped into a small, shallow tray for this purpose, or a fine-mesh tea strainer can be used. For small quantities a shaker can be made by stretching a piece of nylon mesh fabric across the mouth of the container holding the enamel powder, fastening the fabric taught with a rubber band.

To apply overglazes
Overglazes are applied over a prefired enamel grounding. The overglaze pigments are prepared by grinding up the finely powdered colour with a painting medium. The grinding has to be prolonged and thorough, the pigments being worked together with the medium on a piece of plate glass with a muller or palette knife. A muller is a flat-bottomed glass pestle (obtainable from ceramic suppliers). A small muller can be made in the workshop from a solid glass stopper or a thick-bottomed glass bottle: the

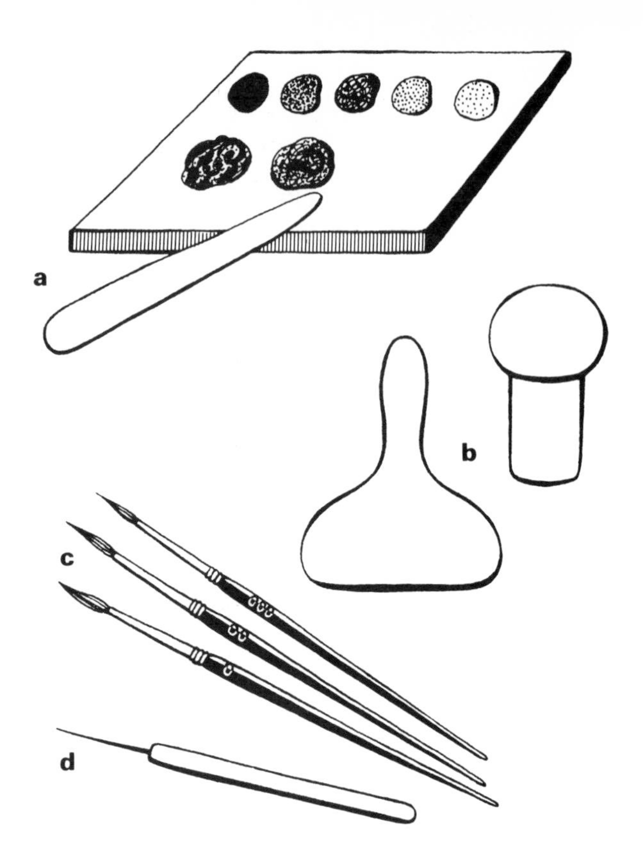

a *A glass slab with prepared painting enamels*
b *Glass pestles (mullers)*
c *Fine water-colour paint brushes*
d *A needle in a holder, for cross-hatching (scratching through unfired glaze)*

base has to be ground smooth with a flat carborundum stone, working under water, the surface left slightly matt.

A set of very fine sable water-colour brushes will be required for pictorial work, from size 000 upwards. For shading and stippling (dotting) slightly thicker, flatter brushes can be employed. A sharp needle set in a holder is useful for cross-hatching. Mapping pens with fine nibs are useful for outlines. Brushes and glass tools are cleaned with commercial paint-brush cleaners.

Metallic lustres can be applied as overglazes during the last stages of the work. Special paint brushes should be reserved for the lustres.

Cleaning the metal surface before enamelling

Acids which can be used for cleaning include sulphuric and nitric acids, but being corrosive they have to be used with great care in a suitable environment. Using strong acids is the quickest method of cleaning the

Miniature ewer
Fine overglaze painting on enamel over a copper base, with the handle worked in relief and enamelled.
Austrian. About 1840.

Cloisonné buttons
By Jane Short. 1979

Two box lids
Monochrome and polychrome enamels over silver with gold and silver foil spangles and a high gloss finish.

Plique-à-jour plate with filigree wires
About 1880. Courtesy of the Bournemouth Museums collections.

metal surfaces but they are not essential. Proprietory solutions are obtainable which take the place of these acids, or a gentle pickle can be made vith vinegar and salt, or the surfaces can be cleaned with detergents and abrasives alone.

The following equipment will be needed for various stages of cleaning the metal before and during the enamelling process: medium and fine-grained carborundum sticks, a brass scratch brush, a stiff bristle brush, a hard toothbrush, a glass brush (a jeweller's tool consisting of a bundle of thin glass fibres), fine steel wool, emery paper (the type which can be used wet or dry), pumice powder, detergent, smooth clean cloth and paper.

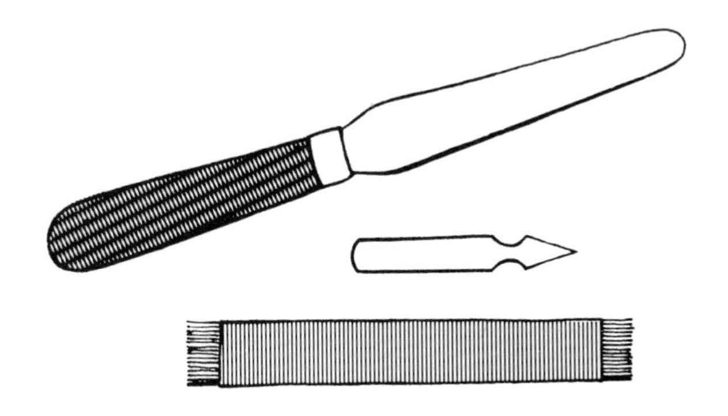

Painting spatula, pen nib and glass brush.

SUMMARY OF EQUIPMENT, MATERIALS, ACCESSORIES AND TOOLS FOR ENAMELLING

1 Enamelling kiln or blowtorch.

2 Firing tools. Firing glove. Firing supports. Cooling slab. Weight.

3 Enamels – stock of various colours and types. Mortars and pestles. Enamelling gum. Storage jars with lids.

4 Pointed tools or brushes for application of enamel pastes. Fine mesh sieves for application of dry enamels. Set of shallow dishes with covers. Spray or atomizer.

5 Overglazes – stock of various colours for painted enamels. Painting medium and thinner. Set of very fine painting brushes. Brush cleaner. Glass pestle (muller) or stainless steel (or ivory) palette knife. Glass slab or glazed tile for grinding and mixing paints.

6 Materials for cleaning metal: acid or vinegar. Brass tongs. Steel wool, wet and dry emery paper, pumice powder, stiff bristle brushes. Jewellers' glass brush. Lintless cloth. Acid-proof dish with lid (if acid is to be used).

7 For filing smooth the fired enamel surface and finishing: medium and fine carborundum stones, graver (for removal of specks), fine steel files, wet and dry emery paper, pumice powder.

8 For mechanical or hand polishing of enamel surface: electric polishing machine with muslin buffs, or polishing sticks. Fine pumice powder. Tripoli. Jewellers' rouge.
9 For etching of metal: acid, acid-resist varnish and/or wax, acid-proof dish with lid, brass tongs. Solvent for varnish.
10 Sheet metal, metal shapes and stock of blanks and/or metal-working or jewellery-making equipment if engraved, soldered, cast or intricately shaped grounds are to be prepared in the workshop. For soldered cloisonné or filigree designs: hard silver or gold solder. Wire. Vice. Drawplates and tongs for drawing wire.
11 For special effects: gold or silver foil. Metallic lustres. Millefiore beads. Enamel threads. Transfers. Scrolling tool.
12 General: scalpel or hobby knife. Sharp scissors. Tweezers. Red felt tip pen or red carbon paper (for outlining or tracing). Sand-filled leather cushion or resilient pad (for grinding and for metal preparation). Tracing paper. Jewellers' pliers, jewellers' saw, metal cutters, raw hide or wooden mallet, planishing hammer. Clean, lintless rags. Supply of clean paper.

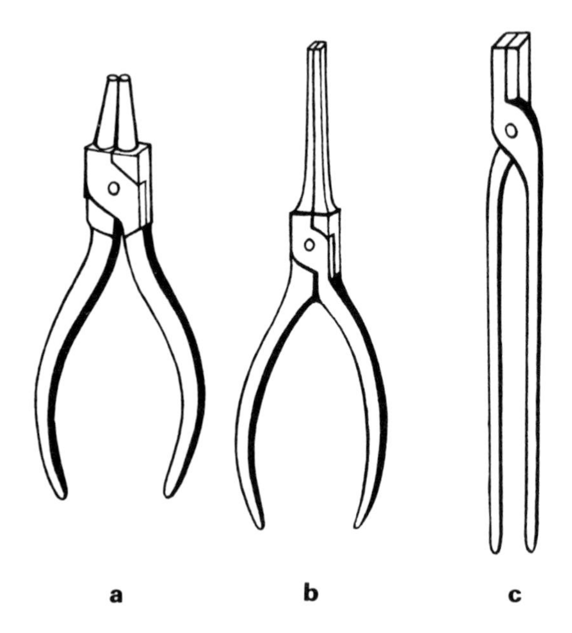

a *Round-nosed pliers*
b *Long-nosed pliers*
c *Heavy close-mouthed tongs for drawing wire through drawplates.*

THE HOME WORKSHOP

If you are setting up a home workshop you will require the following:
1 An area with access to a sink and a rigid table or bench for cleaning of

metal, grinding of enamels, a safe area for working with acids with ventilation or a fume cupboard.

2 A firing area, where the kiln is sited on a strong table, bench or shelf, with sufficient space for firing tools, work waiting for firing and a space where pieces coming out of the kiln can stand to cool. The kiln can be sited in a darker position as this will help in watching the firing, and there should be no draughts.

3 A working area where enamels are applied – this should have good light and dust and draughts have to be avoided. This area should be well away from the 'cooling off' slab to avoid scale, etc. floating into the enamels. Some shelves or drawers nearby will be needed to hold the stock of enamels, overglazes, wires, foils, gum and enamelling tools. Clean, glass household jars are very suitable for storing enamels. One or two display shelves for finished work and a 'notice board' for sketches can be added if desired.

4 Jeweller's bench, metalworking area and equipment – provision for various types of metalworking, such as raising, forming, planishing, engraving, chasing, soldering, sawing and cutting may be needed for certain of the techniques. If the finished work is to be polished with a powered lathe, or polisher, this equipment has to be sited away from the enamelling bench.

Safety precautions

Check that the kiln and other electrically powered equipment is sited away from the areas where water or acids will be used. All electric cables should be protected and channelled away from work areas. Check that the kiln tools have insulated handles (otherwise wind insulating tape around the grips). The flooring in front of the kiln area must be non-conductive and fire-resistant.

Ensure that there is good ventilation or use a fume cupboard when strong acids are used. When diluting acids ALWAYS ADD ACID TO A CONTAINER OF WATER. Dispose of spent acids and pickles safely by neutralizing or greatly diluting them. All toxic materials and containers used with them have to be clearly labelled and stored in a locked cupboard. Wear protective overalls made of cotton (not man-made fibres) when firing or using acids. Avoid open-type shoes when firing.

Tinted goggles will reduce eye-strain when undertaking longer periods of firing.

For sensitive and very young people, protective industrial masks and goggles are advisable when working with dry powdered enamels and when using acids.

DO NOT use hydrofluoric acid in the home workshop as it can cause irreversible injury. (This corrosive acid which dissolves glass is used commercially to remove enamel from precious metals.)

2 Selling your Work

Enamelled articles can compete for custom on many different levels. Costly work can include sets of chessmen, composite pieces such as candelabra, presentation or sporting trophies and also exceptional pieces of jewellery. Traditional articles are in steady demand and include box lids, devotional pieces, badges, lockets, tableware and jewellery with popular designs and colours. Naturally there is interest in modern work and this wide range can include personal jewellery, clock faces or wall and table decorations with brilliant colours. Miniature pieces have collectable interest, particularly for items such as doll's house furniture or models of old artefacts. Small charms can be made with popular appeal, such as animal motifs, emblems of famous cars or symbols of special events. Practical work includes small bowls, containers, holders for specimen plants, menu holders, napkin rings, decanter or bottle labels, match box covers and coasters for glasses. Larger pieces can include picture or mirror frames, door number plates and push-pads for swing-doors, wall plaques with narrative scenes or abstract colour schemes, and series of tiles for mural or table-top decoration. In addition there are many jewellery items which can be enamelled, such as pendants, bracelets, hair ornaments, necklaces, rings, buttons, buckles, brooches and ear-rings.

Attractive, distinctively handmade or professionally finished work will be acceptable for display in local shops and in the gift departments of city stores, providing the price level and style of work fits in with the range of the particular retailer. Certain shops handle only highly priced, traditionally designed articles, some look for *avant garde* ideas and others want only small, cheaper goods with popular appeal which will have a quick turnover. With small shops an appointment to show samples should be arranged with the owner and with larger stores contact should be arranged with the buyer of the department. Single items of work can most readily be sold through art galleries or craft fairs and they can be included in local art exhibitions to encourage commissioned work. Articles offered to shops should be ready to sell – pendants with chains, plaques which are mounted ready to hang, door number plates supplied with fixing screws and small items complete with gift boxes.

Many shops accept suitable work on a sale or return basis. The shop will display your work and pay for items which are sold, returning unsold pieces after an agreed period. When approaching the buyer a few samples of each type of work should be shown together with a price list which gives

Selection of small decorative enamels.
Top row: *Ring, translucent turquoise enamel over silver, and butterfly, with translucent enamels over silver foil, on a copper gilt base.*
Middle row: *Miniature silver screen, with plique-à-jour, and maple-leaf design brooch (silver base).*
Bottom row: *Brooch with enamel and overglaze painting, and brooch with rough-textured enamel and gold lustre with glass bead inset, both on copper bases.*

Examples of traditional enamelwork (early twentieth century).

details of variations which can be made, the delivery times, the unit price and VAT if applicable. The unit price should include all your costs and is compiled by adding the cost of the raw materials, the cost of your working time and a proportion of the overhead expenses (allowances for heating, rent, telephone and postal charges and upkeep of equipment).

For goods supplied on a sale or return basis or for exhibition at a gallery, a receipt for the articles should be signed by the recipient, and this note, which is kept by the supplier, should set out all relevant details of the agreement (including who will be responsible for insurance against loss). If a firm order is given by the buyer of a shop, the accepted procedure is to make out an order confirmation (in duplicate) setting out the details, which is signed by the buyer and the supplier retains the top copy of this on his or her file. When the delivery is made the accompanying invoice should quote the date and reference number of the order so that the customer can check everything. A small discount is generally offered for prompt payment of invoices. If payment has not been made by the end of the month following delivery a statement can be sent. For larger orders it is more practicable to arrange to deliver in two or more small batches spaced over several weeks and each delivery can be invoiced individually.

It is important to keep records of your business transactions. A subdivided file will be needed to hold correspondence with customers, suppliers' catalogues, correspondence relating to invoices received and payments made, and with a 'pending' section for outstanding matters. A large-format address book will save time and trouble and should contain information on suppliers, customers, sources of information, local tax, insurance and VAT offices, bank and service firms. Although very small businesses do not have to register for VAT it is recommended that as a sole trader or partnership a check is made with the local VAT offices to see whether registration is necessary, in which case the VAT paid on goods bought for your business can be offset against the selling price. In this situation you will be advised on what accounting books will have to be kept as a result of registration. A business owner has to keep detailed annual account books for taxation purposes. Advice can be obtained from a professional accountant, or the Small Business Tax Service, or the Business Advisory Service offered by the major banks. Your local branch manager will be able to put you in touch with the department necessary. For National Insurance payments information is obtainable from the Department of Health and Social Security. For information on VAT (Value Added Tax) the Customs and Excise Department has to be contacted. All these Offices have local branches. If you remain self-employed annual accounts have to be prepared. There are various rules for new businesses and special provisions. It is also advisable to check that your insurance policy does not require a special clause to cover the setting up of a workshop if this is on domestic premises.

3 The Techniques of Enamelling

The initial cleaning process of the metal base, the preparation of the enamels and the requirements of firing are very similar for all the techniques.

SMOOTHING THE EDGES AND RIMS OF THE METAL BASE

If the edges of the metal base are left rough or burred this can encourage the formation of fine cracks in the glaze due to tension at these points. The top surface of the metal should be filed smooth to remove any roughness. If the reverse or counter-side of the metal will not be seen in the final article, then a fine burr may be left on this side to help give some hold to the counter-enamel.

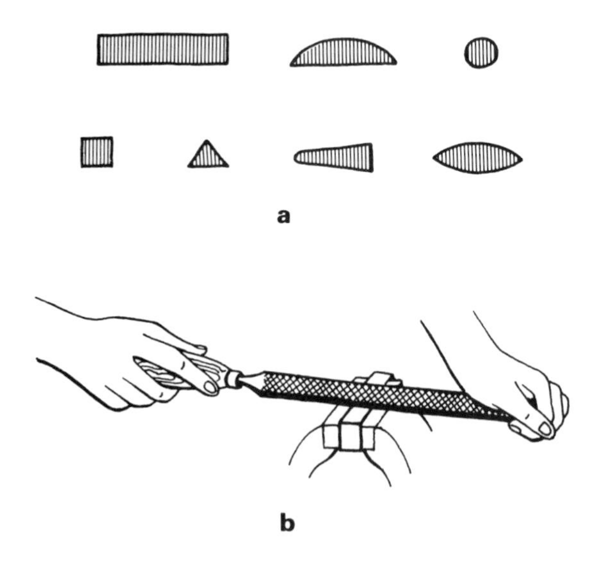

a *File sections*
b *The file cuts on the forward stroke and no pressure is applied on the return stroke, when filing a flat surface the angle of the cut is important*

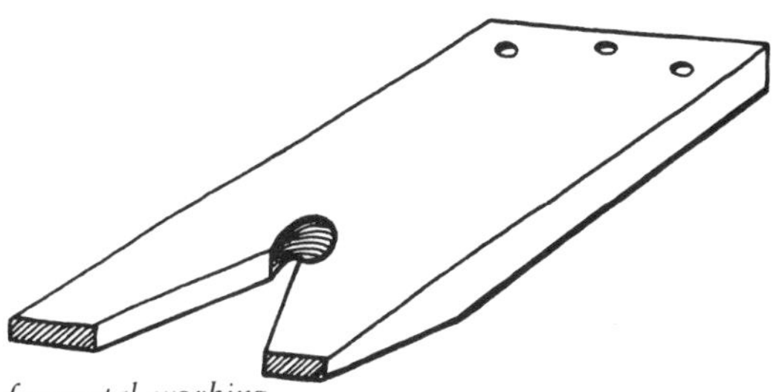

Bench peg for metal working

ANNEALING

Shaping the metal by hammering or spinning hardens it. Annealing the metal prior to enamelling will remove tension and this will help the enamel to bond smoothly to the surface. To anneal, the piece can be placed on a charcoal brick and heated from below with a gas torch until the metal becomes red hot, or it can be placed into the kiln to heat till the metal just begins to colour. Articles of copper or silver are cooled rapidly by plunging into a bowl of cold water but gold is left to cool slowly.

CLEANING AND BRIGHTENING THE METAL SURFACE

Before enamel is applied, the metal surface is cleaned and degreased. For transparent enamels, copper is brightened to increase the reflective properties of the metal.

The quickest and most effective method of cleaning the metal surface is by using a suitable acid or diluted acid pickle. Alternative methods of cleaning are by using scratch brushes, detergents and abrasives. Articles can be heated to burn off surface grease on the metal.

Cleaning with dilute acid.

Cleaning with acid

A glazed earthenware or Pyrex (heatproof glass), lidded vessel will be needed and copper, brass or wooden tongs must be used to place and remove articles when using acid.

Cleaning of the metal surface is always followed by rinsing and drying immediately. If enamel is not to be applied immediately the metal should be wrapped in tissue paper. Once clean the metal surface should not be touched with the fingers to avoid leaving prints, and the piece should be held by its edges or moved with tweezers or a spatula.

When making up an acid pickle, sulphuric acid is added in the required proportion to the dish already filled with water. It is essential that the acid is added to the water, not vice versa, otherwise a reaction can take place which causes the acid to spit out of the dish. The diluted sulphuric acid solution is generally referred to as pickle. The cleaning action of sulphuric acid pickle is speeded up if the acid bath is warmed over a suitable heat source. Undiluted sulphuric acid and other acids should not be heated. The pickle can be used many times, but gradually loses strength and darkens.

Nitric acid is extremely corrosive, giving off dangerous fumes on contact with metal and it must be used with suitable precautions. When neat nitric acid is used, contact with the metal must be for one or two seconds only and the acid must be thoroughly rinsed off the article immediately, preferably using warm water, following this with a dip into a weak detergent solution.

A solution of two tablespoons of common salt in a beakerful of good quality vinegar will make a harmless pickle suitable for cleaning and brightening copper.

Proprietory cleaning preparations for the different metals are available from craft suppliers.

To clean gold

Make a pickle of 80 to 100 drops of sulphuric acid in one litre (about 1.75 pints) of water. Leave the piece in the pickle until the gold turns yellow. Alternatively, make a solution of one part nitric acid to ten parts of water and leave the article in this solution till the gold turns yellow. Gold can be cleaned without the use of acid by brushing the surface with a glass brush and wet baking powder (sodium bicarbonate).

To clean silver

Make a pickle of 50 to 60 drops of sulphuric acid in one litre (about 1.75 pints) of water and leave the piece in the warm pickle until the surface of the silver appears white.

To clean copper

Make a pickle of 10% sulphuric acid in water, warm if possible, and leave the article immersed until it turns a pinky-brown colour.

Cleaning the copper surface with steel wool, scouring powder and water

To clean heavily tarnished or very dirty copper, and to improve a scratched copper surface, first scour with pumice powder and steel wool, rinse and dip the article into a nitric acid bath for one to two seconds, following this with copious rinsing and neutralizing in detergent water, before drying the surface. Very dirty copper may require a second or third cleaning operation.

Proprietory metal cleaning solutions are offered by some craft suppliers and these should be used according to the directions given by the product makers. Abrasive sponges can be used to scour the surface of the metal. Another method is to clean the metal by immersing in hot detergent water and then scouring with fine steel wool and pumice powder (or proprietory kitchen scouring powders such as Vim). Small pieces can be effectively cleaned by working over the surface carefully with a glass brush and hot water.

APPLYING ENAMEL TO THE PREPARED METAL SURFACE

For controlled effects, powdered enamel, in paste or dry form, is applied to the prepared metal in thin layers, building up the final thickness over two or more separately applied and fired layers. By keeping the total thickness of the enamel quite thin, it is possible to produce the widest range of enamel colours and the fired glaze will appear more refined and brilliant. There is also better adherence to the metal when the glazing is thin. Thick layers of enamel reduce transparent effects and the surface is left more undulating after firing. The thicker enamel glazing tends to crack on cooling and sections may splinter off.

For the inlaid methods the enamels are laid in as pastes, which consist of finely ground enamel and water. To help adherence on shaped work, or on steeply sided vessels and edges of larger pieces, a little gum is generally added to the paste. The traditional enamelling gum is Gum Tragacanth. This is made up in the workshop by grinding up the dry flakes of tragacanth with warm water (it speeds the process if the flakes are first ground up in a small quantity of methylated spirits) and more water is then added to produce a thin gum which feels slightly tacky when tested between thumb and finger. Such a gum will keep for several weeks in a screw top jar and by adding one or two drops of detergent and two drops of household disinfectant to the mixture it will spread more easily and will keep for longer. A very good modern substitute can be made up with cellulose adhesive (wallpaper paste), such as Polycell granules, by sprinkling about a quarter teaspoonful of this powder into a jar of cold water and stirring until the mixture is clear, adding more water if necessary to make a thin gum. Ready-mixed enamelling gums can be obtained from craft suppliers. The gum used with enamels must be able to dry out completely and leave no deposit which might affect the firing. Gum which is too thick may result in bubbles in the fired surface.

Applying enamel paste – the inlaying method

The paste is made by giving finely powdered enamel several rinses of water, the excess water being poured off. The paste is placed into a small, shallow dish and this is supported in a tilted position, with the moisture collecting in the lower half. This allows paste of the right degree of moistness to be taken up. The paste should be reasonably moist for inlaying so that it can be applied easily and evenly. If the paste is too dry it will be difficult to manipulate into the design. Paste which is too wet will tend to overflow the cells and the powder will be unevenly distributed. A

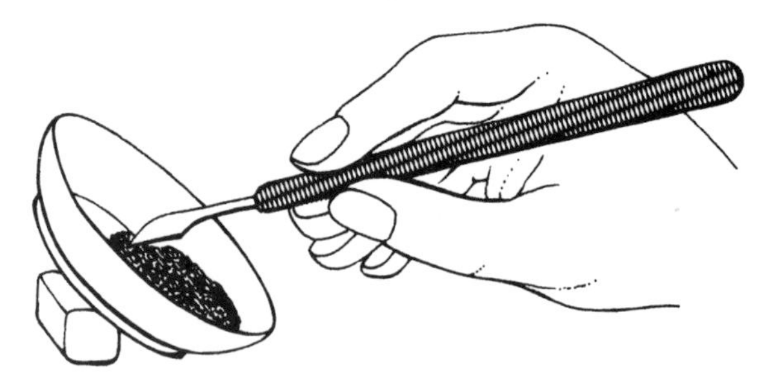

Enamel paste is taken up with a fine spatula or pointed tool from a dish supported in a tilted position

dropper bottle filled with water should be kept nearby and a few drops can be added to the paste from time to time to prevent it drying out.

The paste is inlaid into the design with a pointed quill or with a fine

Inlaying enamel into a champlevé design, using a quill.

spatula or point. As each section is filled with paste the side of the article is tapped gently with e.g., the wooden handle of the spatula. This helps the paste to settle into a smooth layer and brings trapped air and residual moisture to the surface. Excess water is blotted off by holding a small pad of linen or absorbent paper to the edge of the enamelled area. A painting spatula can be used to apply and spread a stiffer paste over a larger, flat metal surface.

Dry application

To apply enamel as a dry, fine powder, the metal surface is usually first coated with gum, so that the dusted-on enamel will not become displaced when inserting the work into the kiln.

To cover larger areas, the metal surface is coated with a thin gum, which can be sprayed on or painted on with a soft brush. To cover shaped work or dishes with steeply raised sides, a thicker gum is required which has to be painted on as it will not diffuse in a spray. The enamel powder is filled into a fine mesh sieve or strainer and this is passed over the metal surface, holding the strainer eight to ten centimetres (about $3\frac{1}{2}$ in) above the surface and tapping the side of the strainer with a finger to deposit a fine, even layer of enamel over the surface.

Dry enamel powder can be trailed across the gum-coated surface or deposited in specific areas by folding a piece of card into a V-shape and tapping the powder down through the folded point of the V.

Applying Counter-enamel

The counter-enamel (on the reverse side of the metal) is applied either by dry application or by spreading paste over the whole surface. When dry,

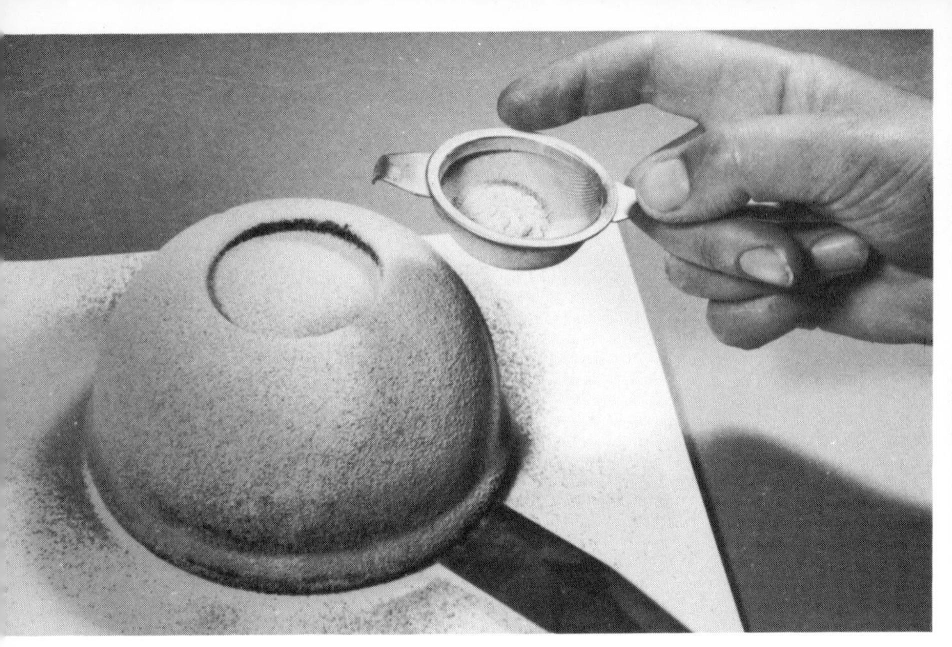

Applying dry, finely powdered enamel onto the gum-coated metal surface with a fine-mesh sieve.

the piece is briefly fired, to fix the counter-enamel, preferably with the counter-side facing upwards.

APPLYING COLOUR FOR PAINTED ENAMELS

For painted work the enamels or metallic oxides are applied over a prefired enamel ground. The painted work is applied with fine brushes and/or with points, the design being built up over several separately fired layers.

FIRING

Firing allows the applied layer of enamel to fuse into a smooth glaze which adheres to the base. Each piece of enamelwork generally requires several firings to complete the design. The length of time which is required to fuse the enamel to the metal and to fuse the upper layers to the grounding varies according to the thickness and overall size of the metal base. The different metals also vary in the speed with which they heat up. Another important factor will be the stage which the work has reached. The dimensions of the muffle and the positioning of the work inside it will also create specific conditions.

Work is always placed on a support for firing. A stilt, or a U-shaped, or W-shaped cradle will be necessary to suspend counter-enamelled work to prevent adhesion to the support. For larger pieces the temperature of the muffle is raised sufficiently to allow for heat loss when the article is inserted. The work is inserted at working temperature and withdrawn as soon as the enamel has fused. Small pieces of thin metal may require less than one minute to fire. Larger pieces an average of one to two minutes.

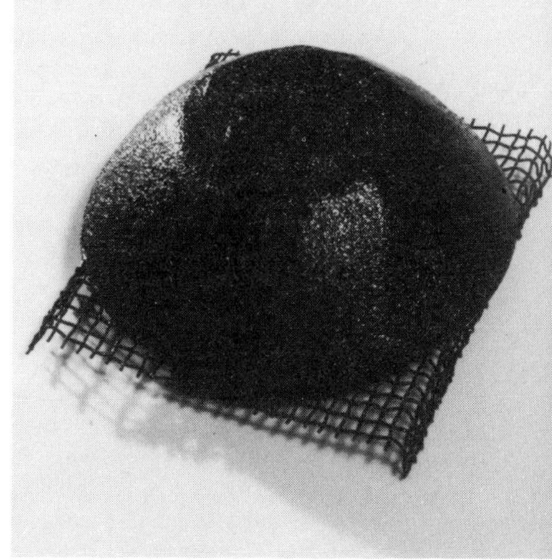

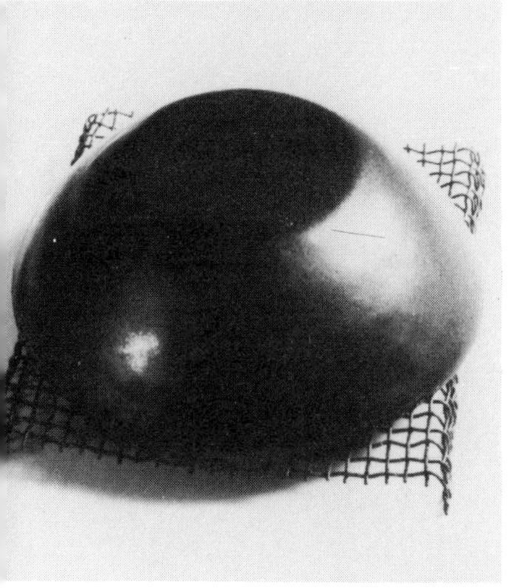

The bowl is covered evenly with enamel, placed on its support to dry completely, then warmed in front of the kiln before firing.

Underfired glaze with a dimpled or 'orange-peel' surface.

Firing to a high gloss.

It is important to dry work completely before firing. If any residual moisture is present during firing this will start to boil and causes grains of enamel to become dislodged. Work is left to dry on top of the hot kiln or it is held on its support in front of the open muffle, until no steam rises from the surface. The dry enamel appears pale and powdery. When ready for firing the support is held with the firing tool and inserted into the muffle and placed gently on the muffle floor as any jarring could dislodge the dry enamel. The firing tool is withdrawn and the kiln door is closed. (For some of the overglazes the door is left open for a few seconds to allow the escape of vapours.) The firing is monitored by opening the kiln door from time to time or by watching through the inspection hole in the door. The muffle tends to darken at first due to heat loss. As the temperature builds up again the metal heats and starts to glow and at the same time the enamel begins to melt. Some colours spread out and flatten very quickly, others first shrink into tiny globules before smoothing into a glaze. Just before flowing into a smooth coating, the enamel reaches a dimpled or orange-peel stage. As soon as the enamel has fused correctly, forming a fluid glaze over the metal, the work is withdrawn, on its support, and placed near the warmth of the kiln to cool. If it is difficult to be sure whether the enamel has fused a long metal spatula can be held just above the surface of the work in the kiln and if the enamel has fused it will be possible to see the reflection of the spatula in the molten surface. However, it is safer to check whether firing is complete by partially withdrawing the work and replacing it in the muffle if it is still underfired. When inspecting the surface the work should not be tilted, to prevent the enamel flowing. If the muffle is not uniformly hot, or when very large plaques are being fired, a more evenly fired surface is produced if the work, on its support, is removed from the muffle before firing is quite complete, rotated 180° and briefly reinserted into the muffle to finish the firing. If copper articles are rotated in this way it is important to reinsert the piece again before fire-scale develops, otherwise cooling and cleaning will be needed before proceeding.

Over-firing results if work is left too long in the kiln or if too high a temperature is reached. The results may be loss of certain colours, burning out of enamel which will leave dark patches or rough edges of darkened colour, blackened edges, flowing of the enamel, sinking down of a

Severe overfiring leaves the enamel with degraded colour and shrinkage occurs, leaving puckered and ridged edges with burnt out patches.

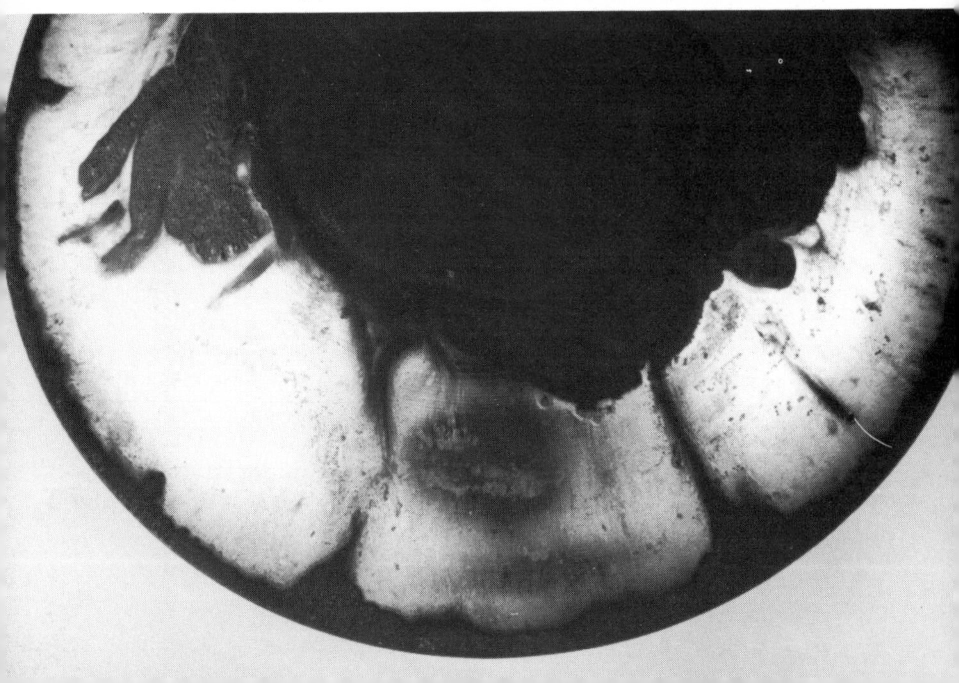

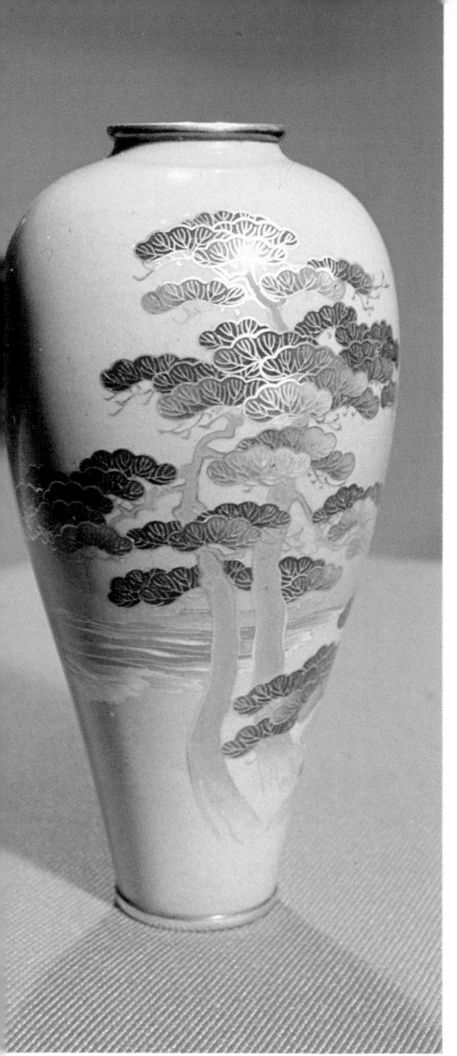

Copper vase decorated with cloisonné Silver cloisons and opaque enamels. By Namikawa Yasayuki. Japanese. Early twentieth century. MCN Antiques, London.

Plique-à-jour cup By Helen M Ibbotson. 1927. Courtesy of The Worshipful Company of Goldsmiths, London.

Small dish and two pendant insets
Transparent dark blue enamel over brightened copper with silver foil motifs.
By Erika Speel. 1981.

Footed cups, made in two sections
Translucent and opaque enamels with liquid gold highlights over copper bases.
By Erika Speel. 1982.

superimposed layer into the grounding and, if grossly overfired, the metal may be damaged.

Underfiring can be rectified by refiring at a higher temperature or for a slightly longer time. Underfired enamel tends to look lustreless and transparents may appear opaque. Very underfired enamel looks granular, pitted or ridged.

Work which will require many firings to complete can be deliberately slightly underfired in the earlier stages of the design to avoid stretching the colours to their limits each time.

Firing work for a slightly longer period at a lesser temperature will not produce as brilliant results as a shorter firing given at the optimum temperature. The aim is to give a short, sharp firing to bring out the full beauty of the enamel colours.

COOLING

Work can be left on its support to cool. Slow cooling is best as sudden cooling can result in cracks forming in the glaze. A piece which requires flattening after firing has to be lifted from the support with a spatula or tweezers and transferred to the cooling slab with a weight placed over the surface. It is possible to flatten quite strongly warped plaques providing the weight is placed in position while the piece is still very hot.

CLEANING AFTER FIRING

Except for the noble metals, firing discolours bare areas of metal. Sterling silver may require pickling and brushing out after firing. Copper produces a dark layer of oxidization or fire-scale. It is important to remove all loose particles of fire-scale because they tend to lift up during refiring of the work and can embed in the molten enamel surface, leaving blemishes which are troublesome to remove. The fire-scale can be removed, when the piece has cooled, with carborundum, steel wool, emery paper or an abrasive sponge. Alternatively, the article is immersed in a 10% sulphuric acid until the discoloured layer has dissolved or can be easily brushed off under water.

STONING

Medium- and fine-grained carborundum stones are used to smooth down or to remove excessive thickness of the fired enamel glaze. This process is known as stoning. The carborundum stone is also used as a file to remove enamel from the edges of the metal or to bring the enamel flush with the metal outlines. Small blisters or blemishes in the glaze can similarly be stoned off. Stoning is always done under water, preferably under running water, to flush away the loosened particles which would otherwise become ground into the enamel surface. Fragile work is supported during the stoning as a certain amount of pressure is needed. Stoning is followed by brushing out under running water with a stiff bristle or glass brush. Stoning the enamel surface leaves a matt finish.

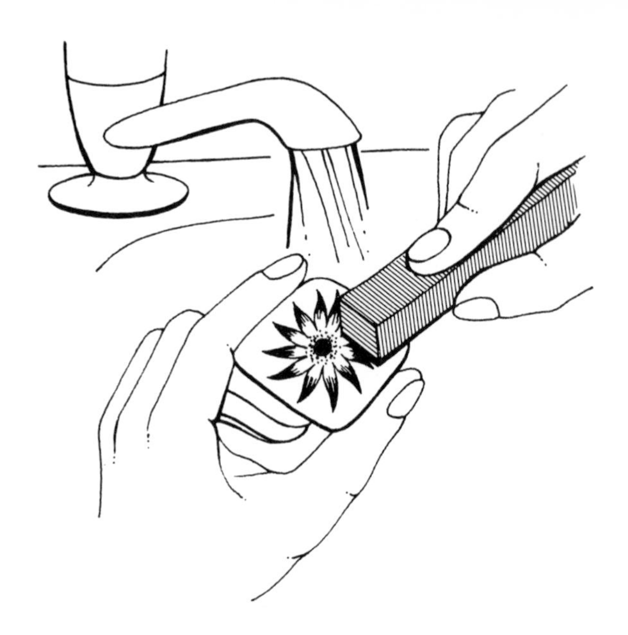

The fired surface is smoothed as necessary by filing with a flat carborundum stone under water

FINISHING

Finishing the fired enamel surface helps to give a smooth and precious appearance to the work and this is important for small-scale jewellery work in traditional champlevé, basse taille and cloisonné. The surface requires stoning after the colours have been fired in to leave it quite level. Then the gloss is restored by a sharp refiring or by polishing.

When the surface is to be polished with an electrically powered polishing machine or lathe, the wheel rotation should be downwards towards the operator. The machine requires safety guards and it should be operable at a slow speed. At the slow speed there is greater control and there is less risk of the enamel cracking. The abrasives for polishing are used as slurries and this work should be carried out away from the other working areas and a protective overall and goggles will be needed. A small broken-in muslin buff is suitable and this is charged with fine-grained, wet pumice powder. As polishing continues more water can be dripped onto the enamel surface with a sponge, so that the surface remains moist. Small or fragile pieces should be attached to a support or embedded in pitch during polishing. The work is held against the polishing buff so that the enamel surface is pressed against the lower part of the wheel, polishing in all directions to avoid striations.

Polishing by hand is a slow process but it gives very good control of the gloss, so that an eggshell sheen or semi-gloss can be given if wanted. The work should be embedded in pitch if it is delicate. The surface is first

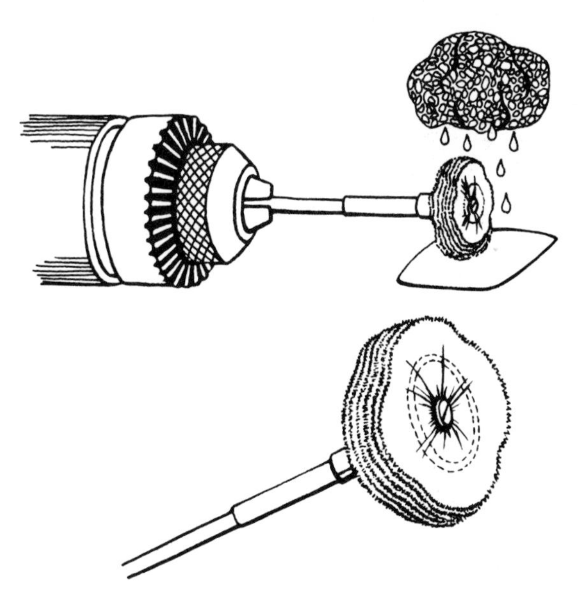

When finishing by polishing with an electrically powered machine: hold the work with both hands, firmly but without excessive pressure against the polishing mop, below the wheel centre. Never put the edge of the article directly to the wheel, but work from the centre of the surface towards the edges.

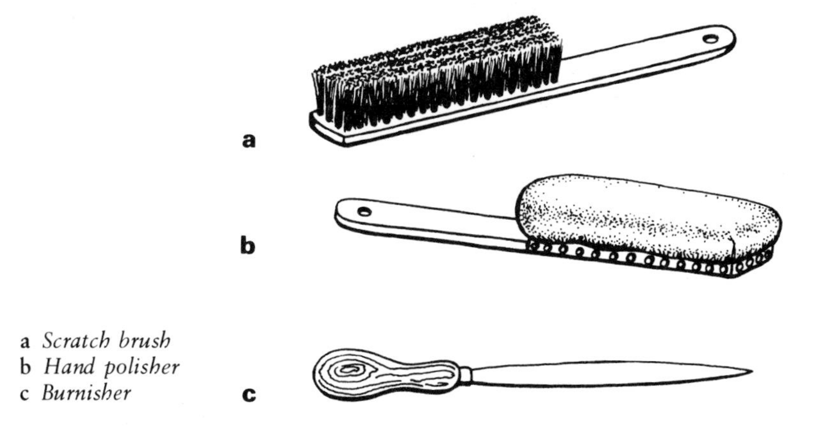

a *Scratch brush*
b *Hand polisher*
c *Burnisher*

rubbed down with the flat side of a very fine-grained carborundum stick or with fine wet and dry emery paper, washing off loosened particles frequently. A series of polishing agents is applied, all used wet and applied on separate polishing rags or felt-covered buffing sticks, removing all traces of the previous powder before applying the next. Fine pumice powder is used first, then rottenstone and putty powder. Jewellers' rouge

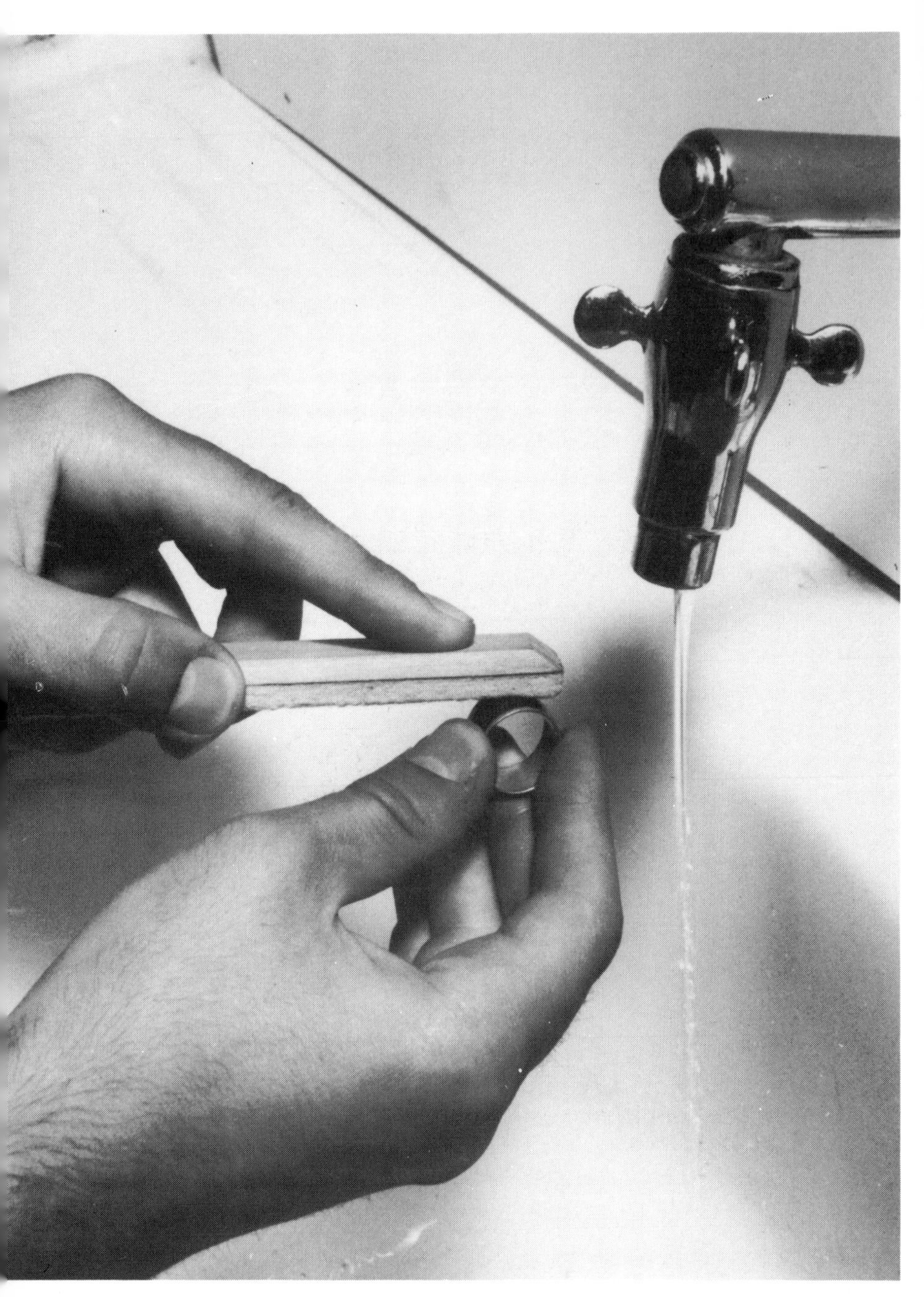

Polishing a ring using a felt-covered stick, pumice powder and water.

can be used to give a high gloss to a fluxed surface, but may cause some staining on other types of enamel.

COLOUR TESTS

Colours should be test fired before use on important work. Each colour can be fired on an individual blank or a larger test plate can be prepared over which several colours can be arranged. The test plate is prepared and fired with different types of grounding: the plate is cleaned and brightened and flux is laid over the central part, leaving a bare strip at each end. After firing and cleaning, the bare strips are covered with opaque white enamel at one end and the colour to be tested at the other end. The central part is covered with strips of silver and gold foil, leaving one strip in flux. After firing the plate is ready to have the colour laid over all the strips, so that when refired the effects of the different groundings can be seen. For opaques the strips of foil can be omitted. The reference number of the colour is noted on the plate.

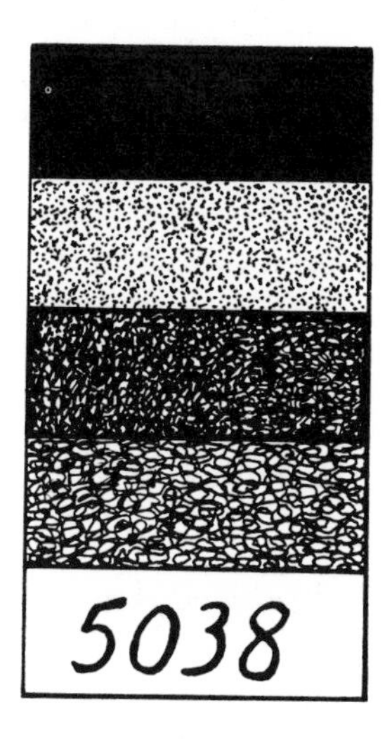

Colour test

RAISED CELL ENAMELLING OR CLOISONNÉ

In cloisonné enamelwork, a raised cell design is made by attaching narrow strips of metal to a base plate to form compartments into which coloured pastes are filled and fired. In traditional cloisonné, the divisions or fences (cloisons) are very thin gold, silver or copper strips or flattened wires, attached vertically (edge-on) to the base so that the top edges of the cloisons show as fine metal threads in the finished surface of the work after the enamels have been infilled. The cloisons give well-defined outlines to the design, separating enamels of different qualities so that they cannot intermingle during firing. The cloisons also act as supports which help to

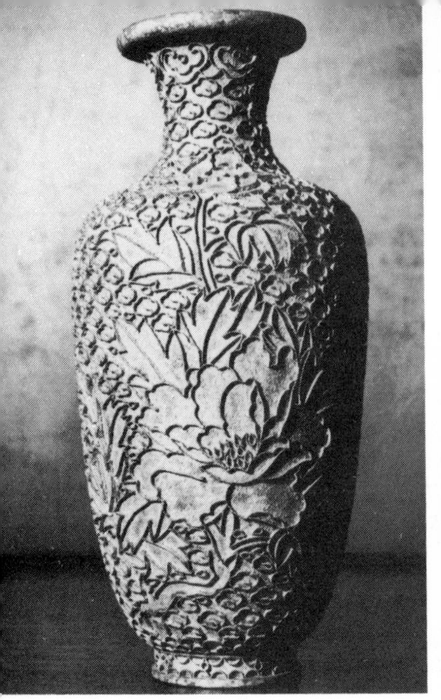

Cloisonné wires soldered to a metal base.

Close up of cloisonné wires.

hold the enamel firmly to the base. The traditional cloisonné method is most suitable for small-scaled work. Modern interpretations of the technique range from highly organized designs to large abstract compositions in which the cloisons give linear interest.

Cloisonné is one of the oldest enamelling techniques. Although known in certain areas in earlier periods, the method was perfected by Byzantine craftsmen who worked on small-scaled articles of gold during the period from the late ninth till the end of the twelfth century A.D. From these Eastern European workshops, the cloisonné technique eventually spread to other areas. In Eastern Europe the true cloisonné method was to be superseded by the simpler filigree (round wire) technique. From the fifteenth century A.D. cloisonné enamelwork was made in China, mainly on large-scaled works of bronze and copper. The method reached Japan in the seventeenth century and flourished there in the late nineteenth and early twentieth centuries, when very intricate articles were made. Oriental work was imported into the West during the second half of the nineteenth century and this brought about a renewal of interest in cloisonné enamels. This coincided with the growing interest in medieval crafts during that period. These revivals led to new developments in cloisonné enamelling with broader styles.

For jewellery purposes, gold and silver are naturally the most suitable and easily worked metals. The cloisonné method allows quite thin metal bases to be used. The cloisons can be cut from thin sheet metal, but generally fine gauge flat wire is used.

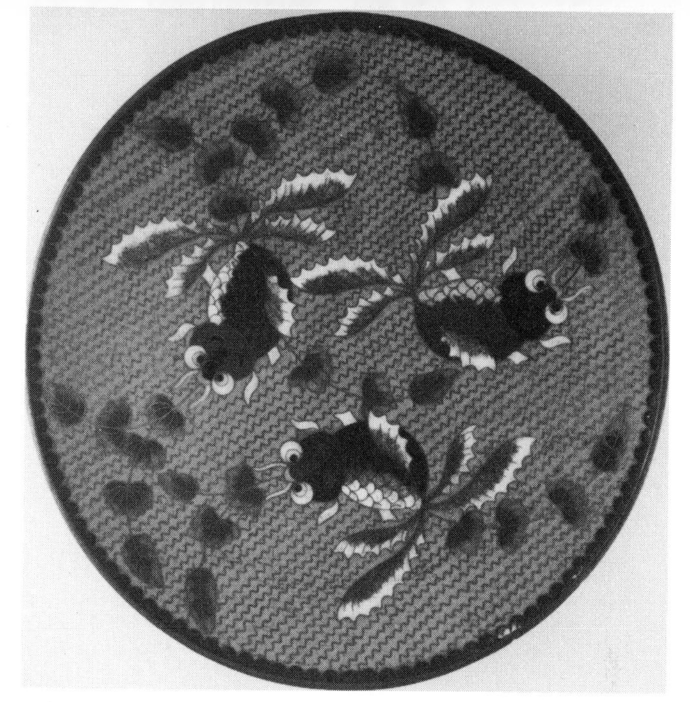

Cloisonné on copper.

The metal base can be a flat plaque or a shaped surface. The cloisons are attached to the base by soldering in position with a high firing solder – either a hard silver solder or gold solder – or they can be attached by being partially embedded into a grounding layer of enamel fired over the base. When the cloisons are soldered to the base, it is usual to turn up the rim of the base plate or to solder on a cloison at the perimeter of the enamelled area, to form a wall which contains the enamel. If part of the surface is to be left without enamel, the base is indented to form recesses into which the cloisons are set, so that in the finished piece the enamelled parts will be flush with the surrounding metal. With silver a thicker base plate is required to allow the cloisonné work to be recessed.

The main outlines of the cloisonné design should be formed with continuous strips when possible to reduce the number of joins. Free-standing straight lines should be avoided as they tend to fall sideways during fixing. Where the wires meet to form cells the joins should be as unobtrusive as possible and cutting the cloisons with a jeweller's saw or with a sharp chisel at a slight angle will give better joins than cutting with scissors. The cloisons can be shaped and contoured to fit the base plate quite easily if the wires are first annealed.

Small-scale work in gold can have a base plate of gauge 24 or 0.56 mm (0.022 in) and for silver and copper a slightly thicker base is used, generally gauge 20 or 0.91 mm (0.036 in). For larger scaled work on copper, gauge 18 or 1.22 mm (0.048 in) is suitable. Counter-enamelling is necessary when working with bases of copper and it makes work on gold more rigid.

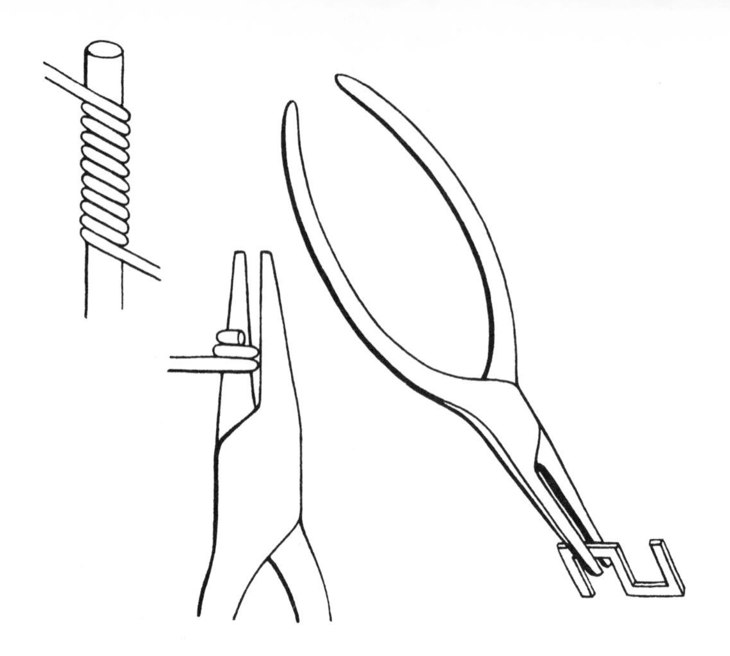

Shaping flat and round wires for cloisonné or filigree

To solder cloisons to the metal base

The cloisons need only to be fixed to the base plate at intervals, or at key points. Enamelling quality (hard silver) solder is essential and only the minimum of solder should be used. Any solder which flows where enamel is to be applied should be filed off. Prepare the strip of solder by hammering it into a thin foil, cutting the end into several narrow strands and cutting horizontally across these to produce tiny chips of solder. The

Cutting hard silver solder into small chips

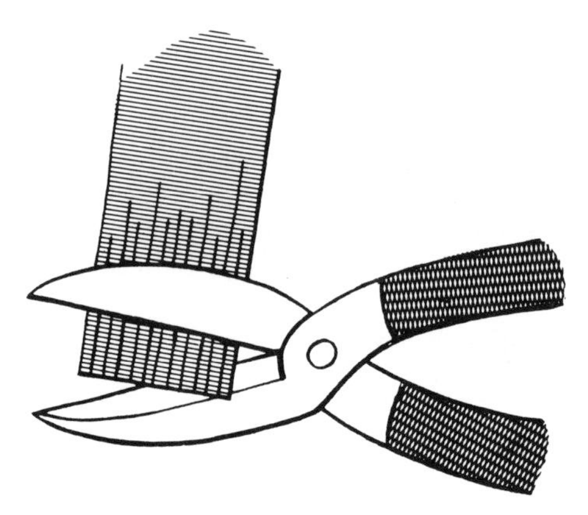

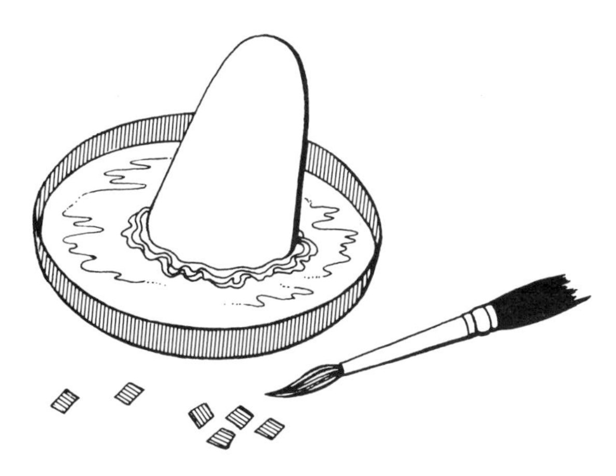

A cone of borax is rubbed in a shallow slate dish with a little water to make a creamy paste for soldering

metal base plate and the cloisons need degreasing before solder is applied. The cloisons are positioned on the base plate according to the design, attached with a little fish glue and if necessary held with binding wire. Using a brush or fine tweezers, individual fragments of solder are picked up, dipped into a thin borax and water paste (which is the fluxing agent which makes the solder flow) and placed at intervals along the bottom edges of the cloisons where they meet the base of the plaque. A little borax paste is painted along the joins to make the solder flow where required. The whole article is heated with a blowtorch, heating the surface of the metal evenly until the solder flows. It is also possible to heat the work by placing it into the muffle, removing it as soon as the solder melts. The article then requires cleaning.

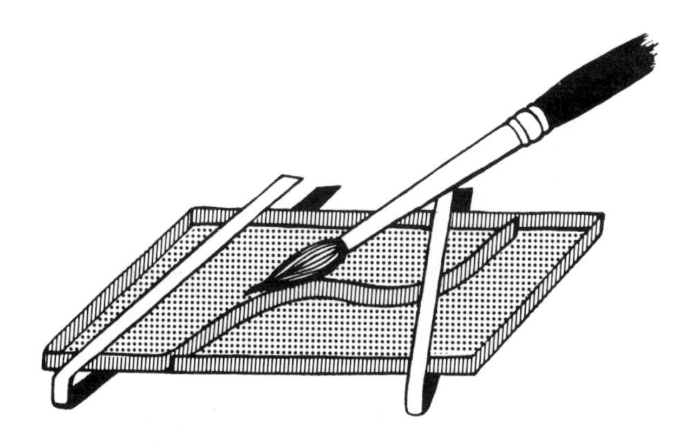

Cloison held to base plate for soldering

Embedding cloisons: Cloisons embedded into the enamel grounding during firing – on removal from the muffle, any cloisons which have not sunk down sufficiently can be gently pressed down while the enamel is still molten.

To fix cloisons by the embedding method

Embedding is a quicker and more direct method of fixing cloisons to a metal base and allows greater flexibility of design as further outlines can be attached in separate stages. This method also avoids the problems which solder tends to produce with enamelwork. There is, however, a little less control than with soldering as the cloisons may become displaced during embedding and very fine wires may sink down too deeply.

The cloisons are prepared and the metal base is fired with counter-enamel and with a grounding of flux. Using flux allows a wider choice of colours for the design, but any hard grounding can be substituted to suit the work. When the grounding has been fired, the surface may require smoothing by filing with carborundum stone, so that everything is level. The grounding is next coated with gum. The cloisons are lifted with tweezers, the bottom edges dipped into gum and positioned on the grounding. Before the gum dries a fine covering of flux is sieved over the surface. When the gum has dried, any grains of flux which have lodged on the top edges of the cloisons are brushed off. Before firing the piece should be well warmed and checked to ensure that nothing has been displaced before inserting into the muffle. As soon as the grounding is seen to melt and the cloisons begin to sink down under their own weight, the piece is withdrawn from the muffle. A spatula should be kept ready by the cooling slab, and any cloisons which have not embedded can be carefully pressed down to the right depth into the still molten grounding.

For simple designs the cloisons can be fixed directly to the metal base with gum. The dry flux is then sieved over the design so that the first firing fuses the grounding to the base with the cloisons embedded in it.

Filling in the colours

The coloured pastes are laid into the cells composing the design in even layers. The pastes should be well tamped down with the inlaying tool to prevent air pockets forming in the glaze. The side of the piece is tapped to settle the enamel. The surface moisture is drawn off with a piece of linen or blotting paper, holding the absorbent pad over the enamel or to the edge of the article. The first layer of paste is not brought up as high as the tops of the cloisons. The second layer is applied to bring the enamel to the required thickness. When this has been fired, a third layer may be needed to even out any areas which have shrunk down and for the third layer the enamels can be filled slightly higher than the cloison tops. After firing the surface is filed smooth to obtain a level surface, using medium and then fine carborundum stone under running water. The gloss is then restored by refiring or by polishing.

Filigree

For filigree the outlines of the design are of fine round, beaded or twisted wires. The filigree outlines can be soldered or embedded in place as for

Filling paste into a cloisonné design.

Excess moisture is absorbed with a folded cloth or blotting paper pad touched to the surface of the paste.

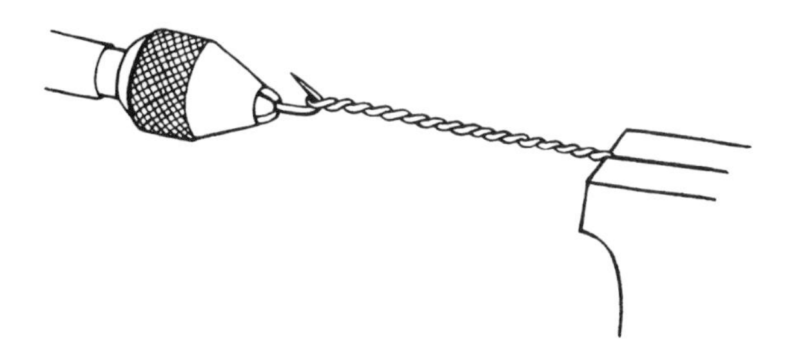

Annealed, cleaned wire can be twisted with a hand drill and vice, for filigree designs

cloisonné work. Some types of filigree work are only partly enamelled, leaving the remainder in plain metal.

With the filigree outlines the enamel tends to fire with a concave or dished effect, due to the shape of the wire outlines which pull the enamel upwards. A single infill of enamel may be sufficient to colour filigree designs. A second layer can be added to give thicker cover and a third layer can be fired if the enamel is to fire with a beaded effect, slightly raised above the metal outlines. Filigree enamel work is not filed or polished.

Filigree enamel.

Simple filigree with beaded wire decorating a pill-box lid.

SUNKEN CELL ENAMELLING OR CHAMPLEVÉ

For champlevé (raised field) enamelling, a sunken or recessed design is worked into the metal surface to create cells or channels into which enamel is filled and fired flush with the metal left at the original level. The metal divisions which are left upstanding as the design is created can be of varying thickness and a greater or lesser proportion of the whole surface can be recessed for enamel. The design can be worked into the metal surface by engraving or other silversmithing techniques.

Champlevé is one of the oldest and most durable of the enamelling techniques. Small-scale examples of champlevé enamelwork made by Celtic and Roman craftsmen in the early centuries of our era still survive. During the medieval period the great goldsmith-enamellers of Western Europe produced outstanding champlevé work, generally on copper plaques which were inset into devotional articles. Limoges was the centre

Copper gilt plaque, formerly filled with enamel: showing how the metal was engraved for champlevé. The recessed cells, engraved to an even depth, were left slightly roughened. The faces were left in metal with engraved lines to show features. (Moses and the Brazen Serpent, twelfth century.)
Courtesy of the Victoria and Albert Museum.

for the largest output of champlevé enamels during the thirteenth century A.D. Examples of champlevé enamelwork dating from various periods can be seen in the collections of most national museums. Champlevé has continued in popularity on heraldic and insignia work and for articles which have to withstand wear and tear, including tableware, writing implements, box lids, buckles and rings.

Designs for champlevé are most effective if they are clear cut and bold. Everything must be planned in detail before commencing the work. However, it is possible to add fine lines or details, after the design has been engraved into the metal, by incorporating cloisons. Opaque colours give the best contrast against the stronger metal outlines of champlevé work and opaques are also more suitable because by being recessed into the metal in narrow channels, transparency tends to be greatly reduced. If transparents are used, they can be improved by firing an initial layer of flux or opaque white in the cells. It is also possible to line the cells with gold or silver foil.

Detailed designs are generally engraved into the metal surface. It is possible to create the recessed design by chasing or embossing, if thin gauge metal is used. Etching techniques are suitable for thicker gauge metal when there is some freedom of design.

To engrave metal for champlevé

Small-scale work can be embedded in pitch during the engraving. Larger pieces are supported on a sand cushion over the bench. For copper the metal base should be at least gauge 14 or 2.03 mm (0.08 in). Generally the depth which is cut away is up to one-third of the thickness of the base plate.

The design is transferred to the metal surface and the outlines scratched in with a scriber. The design is then sunk into the surface with engraving tools to an even depth in all the cells. It is important to avoid undercutting

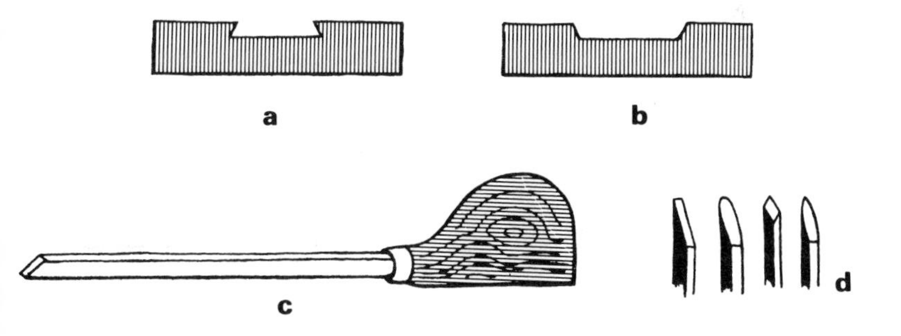

a *Incorrect cut for champlevé (the cells are narrower towards the top)*
b *Correct cut for champlevé*
c *Engraving tool with flattened handle*
d *Graver points*

or an inward slope to the sides of the cells. By leaving the sides of the cells vertical or sloping outwards slightly towards the top, there will be less pressure on the enamel and this reduces the risk of cracks developing. Where possible, larger areas should be subdivided by leaving upstanding divisions to help ease tension on the enamel. The bottoms of the cells or troughs should be left roughened as this will help the enamel to grip, or key into, the metal more securely.

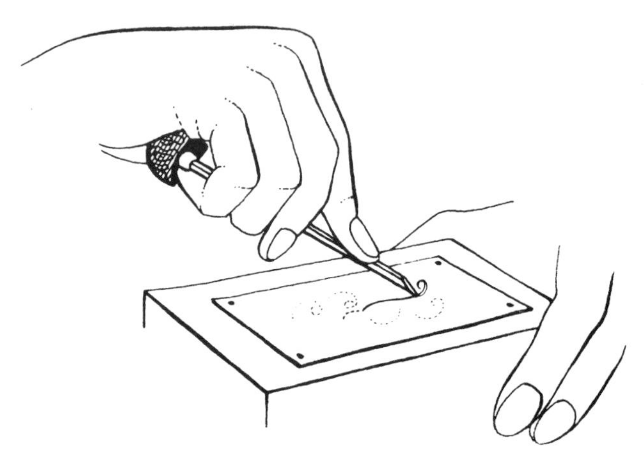

Engraving for enamelling. The tool is given some pressure downwards. The angle of the graver controls the length of cut – a high angle gives a long cut, a lower angle gives a shorter cut. Cutting is to an even depth in all the cells for champlevé and in stepped relief for basse taille.

False champlevé

This method produces a recessed design which can be filled with enamel in the same way as true champlevé but there is no need to use engraving skills. Two flat, matching metal plaques are needed: the top plaque is pierced out with a jeweller's saw to give an enclosed design (no shapes open to the rim) and the fretted plaque is attached to the bottom plaque with enamelling solder, taking care not to let solder flow into the recessed areas. After each firing the piece requires weighting down during cooling to ensure adhesion of the two plaques.

Etching a sunken design into the metal surface

Silver can be etched, using one part nitric acid in four parts of water. Copper is generally used for etching. Gauge 18 or 1.22 mm (0.048 in) thickness of copper is suitable, but a heavier gauge is preferable.

To etch in a design the metal surface has to be covered with an acid-resisting compound where it is not to be bitten in by the acid. The acid-resisting material can be a stopping-out varnish or a wax. A varnish is applied by brush and is therefore most suitable for curved and complex

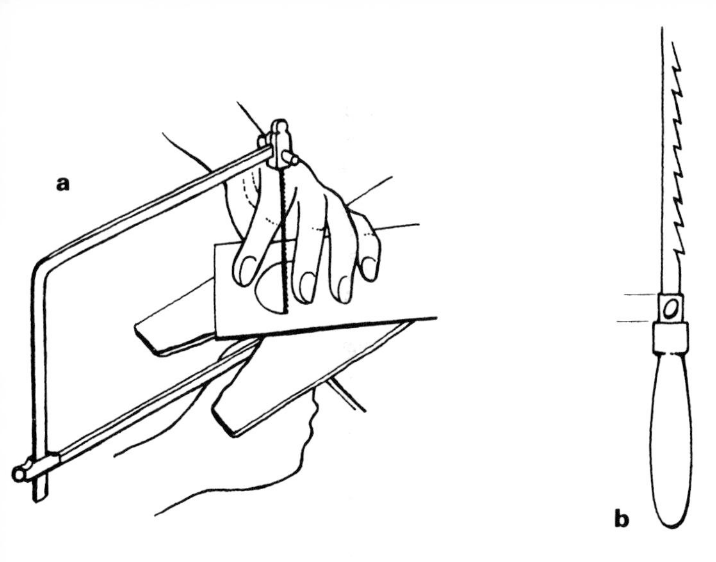

Piercing out sheet metal with a jeweller's saw
a *The metal is supported on the bench peg. The blade cuts on the downstroke.*
b *The teeth of the saw-blade point downwards and they must not be more widely spaced than the thickness of the metal being cut.*

Top plate with pierced out design

The top (pierced) plate is attached to the base plate, to make a recessed design

designs and for touching up parts of the design during the progress of the etching. A waxy resist is very easily applied over the whole metal surface by heating the cleaned metal and floating on the wax. The wax coating can be cut away where required with a point or scalpel to expose the metal and this method is very convenient for geometric designs, straight-sided patterns and fine lines. It is important to give good cover to the edges and reverse sides of the metal plate as well as to the parts to be protected on the top surface. After the etching is complete the varnish type of resist is cleaned off with a solvent and wax can be burnt off.

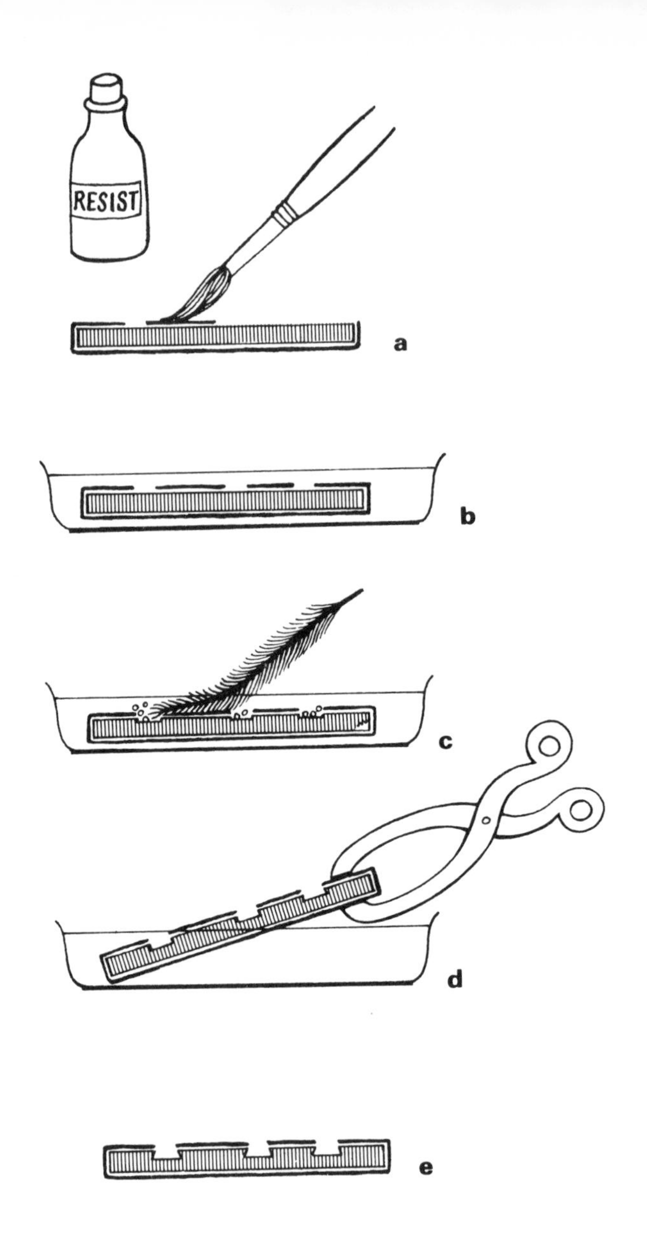

a *All parts of the metal surface which are not to be etched are painted with acid-resisting varnish*
b *The plate is immersed in the acid-bath*
c *Bubbles form during etching and are wiped off with a feather*
d *The plate is removed with brass tongs*
e *Etching tends to undercut, so that the sides of the cells are sloping and the base is wider than the top of each cell*

A copper plate prepared with the resist can be etched in a bath of sulphuric acid, which works fairly slowly, giving good control. For speedier etching, a 30% nitric acid bath can be used. The plate is immersed into the acid bath face upwards. Small bubbles are generated during the etching process and to ensure evenness of bite the bubbles are brushed off the surface periodically with a broad feather. After half to one hour the plate is removed with tongs and rinsed free of acid. The surface should be carefully inspected and where the resist appears thin or worn away it can be patched with a little more varnish. The plate is then returned to the acid bath until a sufficient depth of cut has been reached. There is an unavoidable tendency for the acid to bite in at the edges of the design, resulting in some undercutting. After etching, minor improvements can be made with an engraving tool.

Ferric chloride (which is extremely poisonous) can be used as a mordant, or etching agent, for copper. The process is slower than with sulphuric acid but a straight cut is possible, thus avoiding undercutting. To etch in a bath of ferric chloride the plate is immersed face downwards and supported on slivers of wood to hold it off the bottom of the dish and allow released particles to float off.

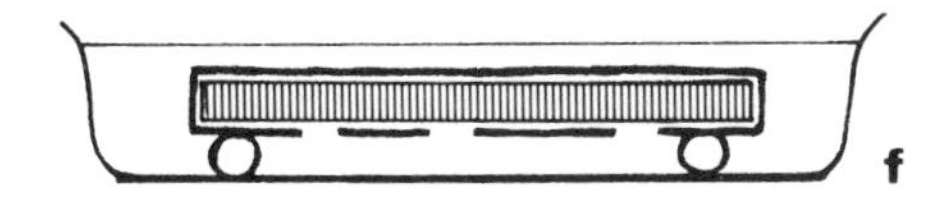

f *To etch in a bath of ferric chloride the plaque is placed face down and supported off the base of the dish.*

When the sunken design has been etched in the metal the piece is rinsed and the resist is removed. Engraving or etching should leave the sunken design clean and ready for enamelling, but if the surface needs cleaning or degreasing this should be done by using alcohol, baking powder, or a detergent solution. Avoid cleaning with acids which could etch away the sharp outlines.

The colours are filled into the recessed design as fine pastes. Each colour is worked into its allotted cells with a point or very fine spatula, pressing well down on the paste to make sure no air pockets are trapped. The side of the plaque is tapped to bring up moisture and settle the paste and as each section is filled it can be dried off with a pad of absorbent cloth. The paste is filled in as a fine layer for the first firing and unless the engraving is very shallow this will not bring the enamel to the level of the surrounding metal at this stage. All the cells are filled for the first firing. After firing, a second layer is applied in the same way, bringing the enamel to the level of the surrounding metal. The firing generally leaves some of the cells underfilled and a third layer is applied, bringing the enamel just above the metal surface. After firing the excess enamel is filed smooth, removing any

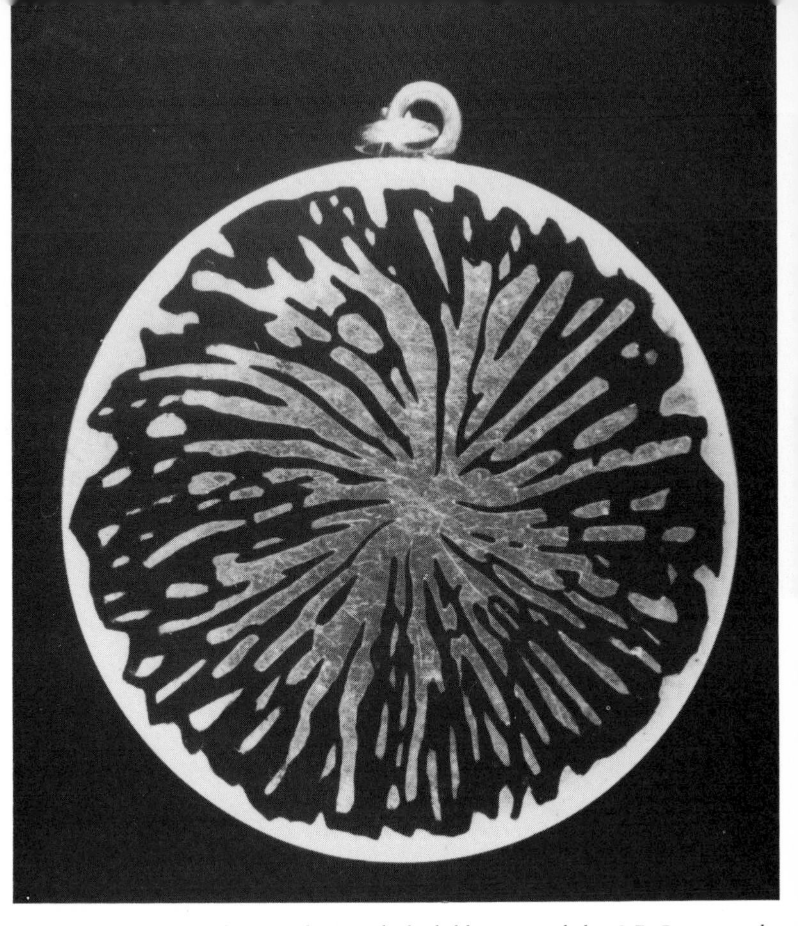

Champlevé design: Silver pendant with dark blue enamel, by C.F. Barnes and A. Mudd.

which may have overflown its outlines, and leaving the enamelled areas perfectly flush with the metal surround and divisions. The gloss is then restored by refiring. The metal areas will require cleaning and polishing and care is required to avoid scratching the enamel.

TRANSLUCENT ENAMELLING OVER LOW RELIEF, OR BASSE TAILLE

For translucent enamelling over a sunken, low relief design worked into metal, or basse taille, the finished appearance is that of a smooth enamel glaze through which an engraved picture or pattern can be clearly seen. Elaborate designs are suitable for small-scale work in silver or gold. For simple and broad designs, copper, copper alloy or copper covered with embossed silver foil can be used. The method can be combined with champlevé.

For basse taille the sunken design is worked into the metal to different

Silver engraved for basse taille, fourteenth century.
Courtesy of the Victoria and Albert Museum.

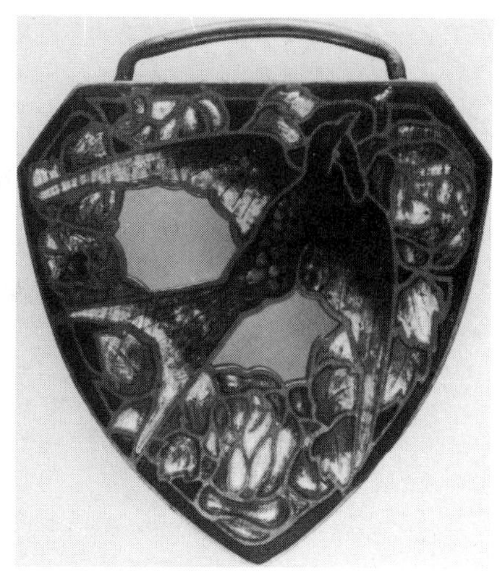

Bird design in basse taille, enamelled with polychrome colours.

levels, the sunken part being enclosed by a rim or edging metal. When the enamel has been fired over the sunken design it forms a smooth coating but there is a greater thickness over the more deeply recessed parts which, therefore, appear as more richly coloured or darker. The higher points of the relief are only thinly covered with enamel, and being comparatively lighter in colour represent the highlights or foreground of the design. The way in which the translucent enamel shades from dark areas to lighter ones enhances the three-dimensional appearance which is produced.

Because of its very reflective surface, silver gives the greatest scope for the basse taille method. For small-scaled and detailed work the design has to be very skilfully engraved into the metal to produce distinct outlines and create the different levels. Regular patterns can be worked into the metal surface by hammering, punching or etching. Metal blanks prepared with a sunken design suitable for basse taille are offered by some craft suppliers.

The basse taille technique was known to the medieval goldsmith-enamellers. Gold or silver plaques enamelled in this way were set into devotional articles such as shrines or plates. Larger pieces could also be embellished with bands of basse taille enamelwork showing narrative scenes. The most notable surviving early, large piece with basse taille enamelwork is The Royal Gold Cup of the Kings of England and France, dating from the fourteenth century A.D., in the collection of the British Museum, and of Burgundian workmanship. During the sixteenth century, when ornate, gem-encrusted jewellery was an important part of aristocratic dress for both men and women, basse taille frequently formed part of these ornaments. The seventeenth and eighteenth centuries saw an adaptation of the method to give geometric rather than pictorial designs and to create backgrounds to painted enamelwork, for example on costly snuff boxes and watch cases. From the mid-eighteenth century, hand engraving was superseded by engine-turning, by which intricate, closely set linear patterns could be cut into the metal by using special lathes. The sharp cut given in this way gave increased translucency to the superimposed enamels and all manner of jewellery articles were decorated with such basse taille. Basse taille in all its variations remained in fashion during the nineteenth and early twentieth centuries and continues to be very popular. The wide range of fine transparent enamels and the easily controlled electrically heated kilns now available have extended the use of the basse taille technique and many decorative and experimental pieces can be classified under this heading now, as well as the traditional work.

The depth of the sunken relief for basse taille is usually in the range of 0.5 mm (0.022 in) to 1.5 mm (0.059 in). For intricate designs and very shallow relief, the best results are generally produced by covering with the lighter transparent enamels or with opalescents. The deeper and richer transparent colours allow less detail to show through and these are at their best over sharply cut designs or strongly contoured geometric patterns.

To work a sunken relief into the metal by engraving

Draw the outlines of the design on the surface of the metal and incise lightly. Remove the metal where necessary to the required depth without any undercutting of outlines and leaving well-defined steps at the different levels. Where possible the base of each cell should be left with a patterned or ridged finish to provide a rough surface for the enamel to key into, so that adherence will be good. Keying in is particularly important when working with silver. On silver it is also advisable to intersperse larger sunken areas with a few divisions left in the metal, for example in figure work the face and hands can be left in metal and the drapery and background can be enamelled, or when decorating a box lid with a geometric design a band of metal can be left at the original level to create a border or subdivide the design into panels.

To work a sunken relief into thin gauge metal

When working a sunken relief design into thin gauge metal, a surrounding edging or lip of metal should be left which forms the perimeter of the enamelled area. The metal plaque is fixed in pitch and the surface is shaped by chasing, embossing or hammering using tracing tools.

Counter-enamelling is advisable when thin gauge metal has been used, to strengthen the work and, prior to setting, the hollows on the reverse of the work should be filled with plaster for added support.

Stamping of the design into metal requires special dies and equipment and it is only practical to use these devices if a very large quantity of articles is to be given the same pattern.

A simple tooled design can be worked into thin gauge metal by placing

Small-scale work is embedded in pitch during metal working, and the illustration shows a piece of work in a Pitch bowl. Underneath is the Collar for the bowl.

Relief designs can be worked into thin metal with chasing tools

the plaque onto a resilient surface and scoring with a smooth-edged tool to leave an indented pattern. Texturing can be produced with chasing tools and a hammer, to give a dimpled or tooled design.

To etch a design for basse taille

The method is basically the same as etching a sunken design into metal for champlevé, using a copper base, letting the acid bite in to different levels by painting in or removing the resist in stages. Where the acid is to bite deeply the metal is therefore exposed for a correspondingly longer period, the more shallow areas being protected with resist as soon as they are sufficiently deep, and details scratched in through the resist as necessary.

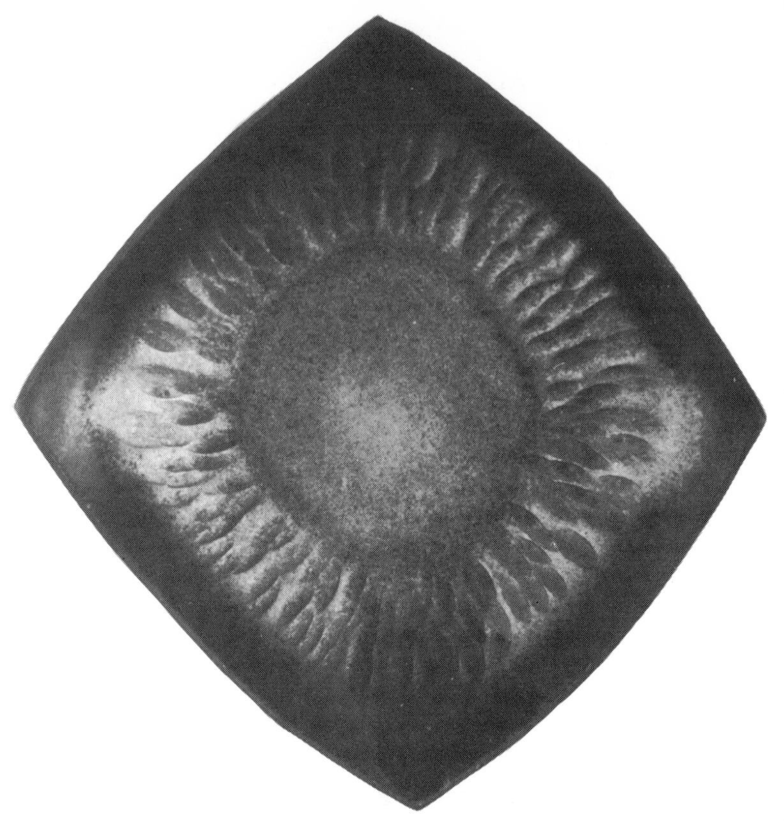

Small copper dish decorated with hammer marks and covered with translucent light blue enamel.

Etching in this way is suitable for broader designs and abstract effects.

When the whole basse taille design is covered with a single colour of translucent enamel the effects of light and shade will be most marked. If contrasting colours are to be used the effects will be of a painting with translucent enamels. Only colours of similar hardness are applied next to each other, otherwise a metal division is left between them.

The colours should be tested individually and selected for maximum translucency over the chosen metal base. Over silver some of the translucents will require a grounding of silver flux. Over a copper base the brightened metal requires a grounding of flux to improve the clarity of the colours which will be applied in the next stage. The first layer of paste should cover the whole of the sunken area but this and each subsequent layer should be kept as fine as possible. If excessive thickness builds up or there are any ridges, these are stoned down before continuing with the work. For basse taille it is particularly important to stone the surface under running water and to scour out loosened particles with the scratch brush. The last layer of paste – which can be of flux to leave a very brilliant, high-gloss finish – should leave the fired enamel just above the level of the surrounding metal or rim. The surface is then stoned to leave a smooth transition from the enamelled parts to the metal areas. The gloss is restored by refiring or by polishing.

ENAMELLING ON HIGH RELIEF, OR ÉMAIL EN RONDE BOSSE

The method of émail en ronde bosse, or enamelling on small-scale designs in gold, silver or copper, worked in the round or in high relief, can highlight delicate goldsmith or silversmith's work and can enhance the visual effect of very tiny ornaments. On very small-scale work the light opaque or semi-opaque enamels produce the best results.

The method became important in Europe during the late medieval period. A large, early fifteenth-century composition known as the Reliquary of the Holy Thorn, in the Waddesdon Bequest, at the British Museum, shows several variations of this technique. In the sixteenth century the en ronde bosse technique was fashionable on costly work, the enamels being combined with large pearls and precious stones on devotional and secular jewellery. Examples of these magnificent Renaissance jewels can be seen in the Jewellery Gallery at the Victoria and Albert Museum, the British Museum and many other national museums. During the eighteenth century the method was adapted for larger scaled work on copper bases, with less delicacy in both the metal working and the enamelling techniques. The en ronde bosse method has been found

Enamel on high relief (émail en ronde bosse).
Courtesy of the Trustees of the British Museum.

particularly suitable for insignia work, jewellery items and on less costly pieces such as hair ornaments, pendants or sporting and club pins.

The goldsmithing techniques of chasing, engraving or casting are generally used to prepare the metal for en ronde bosse enamelling. For small pieces the relief design can be stamped into the metal or gold or silver foil can be burnished over a suitable mould to leave a raised design. For more robust pieces, copper can be stamped to give a raised pattern. For work which is completely in the round, the shape is usually formed from two halves which are soldered together with hard silver solder, a vent hole being left in the base or unseen portion, for the escape of gases during the enamelling. The surface of the metal should be left slightly roughened to help the enamel to stay in place and prevent a flow downwards into the hollows or valleys of the relief during firing and to help permanent adherence. The roughening can be given with the graver, to leave tiny burrs or a zig-zag pattern (the latter being used under transparents) or by scouring with steel wool.

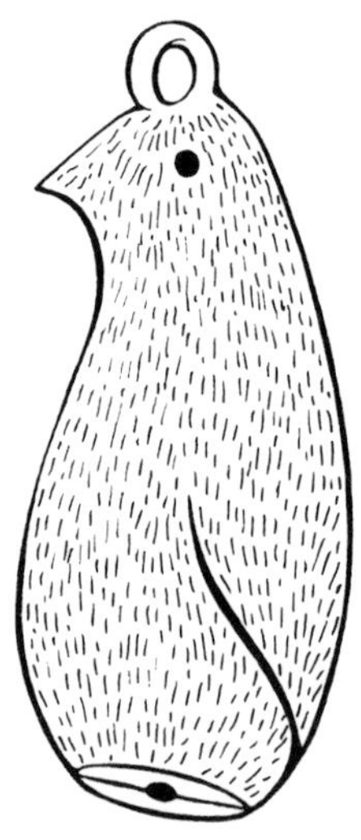

Small hollow silver penguin charm prepared for enamelling in the en ronde bosse method by soldering the two halves with hard silver solder, leaving a vent hole in the base. The surface has been textured with the graver to give a good hold for the enamel.

When a thin metal base is used for the relief work a counter-enamel is desirable. For work which is in the round the counter-enamel can be mixed into a thin slurry with water and gum and blown through the vent hole with a straw, the article being rotated to give even cover of the interior.

When applying paste to the relief design, the point or quill should be used to give thin and even cover over the surface. A little gum added to the paste will help adhesion. The more delicate and intricate the design the more important it becomes to give a very thin covering of enamel to avoid smothering the outlines of the design and loss of delicacy. For larger pieces which require coating with a single colour, cover can be given by painting the surface with gum and dusting dry powdered enamel over, holding the sieve fairly close to the surface to avoid trapping air in the layer of powder. The fewest number of firings possible should be given and two firings may be sufficient to complete the enamelwork. The first firing should be very brief to leave a slightly underfired or orange-peel surface rather than a perfectly smooth glaze, so that the second layer can more easily key to it. If any excess thickness builds up in part of the design this will require filing off with carborundum stone under water, then refiring. Pieces require careful support to avoid marking the enamelled surface. Work in the round may be impaled on a trivet point pushed into the vent hole.

Fine details can be added with overglazes or liquid gold painted on with a fine brush and fixed by refiring.

High relief or en ronde bosse enamelwork is left fire-finished as polishing is not practical.

WINDOW ENAMEL OR PLIQUE-À-JOUR

Plique-à-jour produces the appearance of tiny stained glass windows. The work is made by filling the spaces of an openwork or tracery design in gold, silver or copper with lightly coloured translucent enamels without a metal backing under the enamelled areas. This allows light to pass through the enamel.

Plique-à-jour work was known from quite early times, probably from before the thirteenth century, but it has been too fragile to survive the centuries. One renowned example which still exists dates from the fifteenth century, and this piece, which is known as the Mérode Cup, is in the collection of the Victoria and Albert Museum, London. It has openwork in the form of miniature Gothic windows set into the body of the cup and its conical cover, the enamels set into these areas having little transparency due to the thickness of the inlay and placement. Advances in firing equipment, the improved clarity of transparent enamels and trends in fashion brought interest in the method in the late nineteenth and early twentieth centuries. Such work with very pale translucent colours became popular for jewellery and also on larger pieces such as ladles, demi-tasse spoons, bowls and even lampshades. The technique has become fashionable again in recent times.

The greatest translucency is possible when the openwork spaces are wider in diameter than the depth of the enamel inlay but it is more difficult to fill the larger spaces with enamel due to tensions in the surface which cause cracks to develop. A thin film of enamel gives more translucency but again there is more risk of cracks developing and there are more problems with firing.

The openwork design can be made in several ways. The cells can be formed from shaped wires, as for cloisonné or filigree work, with the cloisons soldered to each other and not to a base plate. Soldering can be avoided by adapting the embedding method: for this a piece of copper or gold foil is cleaned, covered with a layer of flux and the shaped cloisons are placed over this and fired to embed, the work being carefully supported to avoid warping; when the cloisons are fixed the colours are filled into the cells and after firing the foil backing is filed or etched off (the cloisons being protected with acid-resist varnish during etching), to leave the design without a metal backing. A more rigid structure can be produced by piercing or etching out the design from sheet metal.

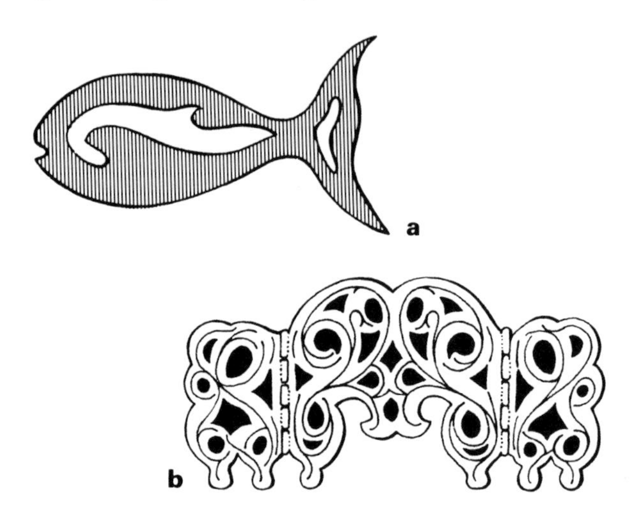

a *Pierced out design for a small plique-à-jour pendant*
b *Assembled sections of a miniature screen with plique-à-jour*

When wire outlines are used, flat or oblong section cloisons will grip the enamel more easily and give a flatter surface than rounded wire. For the rims the wire can be pulled through a drawplate to curve the edges inwards slightly. If the design is made by soldering the cloisons together, solder must be prevented from flowing to the sides which will touch the enamel otherwise there may be problems with adhesion. For pierced or etched designs using thicker gauge metal, the inward facing walls of the cells can be roughened with a graver to provide a better hold for the enamel.

As maximum translucency will be required, the enamel should be well washed just before use. The paste can have one or two drops of gum added, but too much gum tends to produce bubbles in the fired surface.

If the openwork spaces are narrow in at least one dimension the enamel paste may cling adequately to the sides of the cells and firing is possible without the need of a temporary backing to the work. If this is attempted the piece should not be lifted up until the enamel infill is quite dry and held

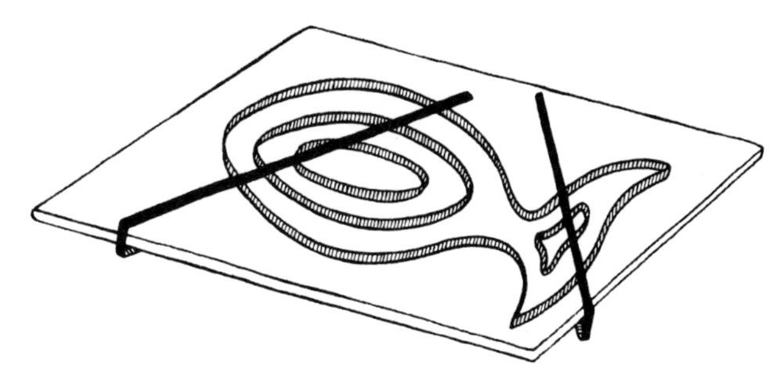

Plique-à-jour: flat work can be attached to a sheet of mica over a support for firing

in place by the gum. When dry, the piece is lifted with a spatula onto a U-shaped support for firing. A temporary support under the enamel is often necessary for larger work. Flat plaques can be supported on a temporary backing of mica or phosphor bronze foil, the work being held to this with binding wire or jeweller's clips in such a way that the enamel is clear of these holds. These backings can be pulled off after firing is complete. If any flakes of mica adhere these can be filed off. Gold foil can be used in the same way but as the foil is finally etched or filed off this is an expensive method. For shaped articles with large openwork spaces the temporary backing is generally gold foil. The traditional method is to make a former (caisson) of metal or fireclay which fits the contours of the article, covering the former with the foil and the piece is attached to this with wires while the enamel is laid in and the work fired. After firing is complete the former is removed and the foil is etched off with aqua regia (a mixture of three parts pure hydrochloric acid with one part nitric acid). Silver foil can be substituted as a temporary backing, providing care is taken not to fire too closely to the melting point of silver. As silver may discolour certain of the enamels it may be necessary to apply a thin layer of silver flux over the foil in the open spaces before filling in the colours. Silver foil is removed by filing. When the temporary backing is removed by etching or filing the matt surface of the enamel on the reverse side can be given a gloss by refiring on a U-shaped support or by polishing carefully.

When laying in the coloured pastes for a plique-à-jour design, each cell is covered to an even depth, blotting each portion dry to help adherence. The cells can be filled almost to the top of the outlines and all the cells are filled for the first firing. As the enamel flows in the firing, there is a tendency for it to pull upwards along the metal outlines of the design, leaving the centres of the cells with a thinner covering. It is quite usual for the work to emerge from the muffle after the first firing with some or all of the cells only partly filmed over with enamel. When this happens it is due to the enamel drawing either towards the outlines in beads, or away from the outlines into the centre of the cell. Where an excess thickness has built up or enamel has flown over the metal due to this, it will require filing down with carborundum. Support the work well during the filing and

Plique-à-jour presentation spoon.

brushing out, to prevent cracks developing. The second and subsequent firing will allow the filling in of holes and the building up of thin areas. A slightly hollowed or concave infill to the cells can be useful as this dishing gives greater translucency to the plique-à-jour. If a flush surface is required, the cells have to be filled slightly higher than the metal outlines for the last application of enamel, and the whole surface is filed smooth after firing. This entails refiring or polishing to restore the gloss. For designs with filigree outlines, the cells can be underfilled, to leave a dished appearance, or overfilled to give a beaded effect after firing. Filigree plique-à-jour is not polished.

When firing and cooling plique-à-jour at any stage, sudden expansion or contraction of the metal should be avoided or sections of the enamel may detach from the outlines. A lengthy warming before firing and a very slow cooling, especially at the finish, on top of the muffle or over a radiator, will help in keeping the work stable and free from cracks.

PAINTED ENAMELS

The painted enamels include several variations of technique. The major divisions are between the methods in which the painted work is created with superimposed layers of coloured enamels and those in which the painted design is laid in with metallic oxides or overglazes. The colours are applied over a grounding of enamel which covers the whole of the metal base. There are no metal barriers to delineate the designs and control of outlines is achieved by the way in which the colours are applied, and with a planned sequence of firing.

Painted enamelwork with monochrome (grisaille), and polychrome transparent enamels over foil underlays and gilding. (Limoges, seventeenth century.) Courtesy of the Victoria and Albert Museum.

The painted techniques were developed in Europe during the second half of the fifteenth century. Limoges, which in earlier times had been renowned for its champlevé enamels, became the centre for the major schools of artist enamellers working in the painted techniques during the sixteenth and early seventeenth centuries. During the course of the seventeenth century, as fashions changed, smaller scaled work created with overglazes superseded the earlier methods. The overglaze or on-enamel painted methods were suitable for elaborate miniature work. Miniature portraits and landscapes were made which could take many hours and up to 20 firings to complete. Other small-scale work was produced more simply with only a few firings and resembling painted ceramic ware which was decorated by similar means. The small-scale painted enamels were produced in quantity during the eighteenth and nineteenth centuries and have continued to be popular. Examples of the various schools and methods of enamel painting can be seen in the collections of many national museums. The Victoria and Albert Museum and the British Museum, London, have major collections of the Limoges schools and of miniature and small-scale painted enamels of various periods.

A counter-enamel is required for painted work when copper or gold is used for the base. Thin gauge metal is more suitable for the painted techniques as there will be less expansion of the base and firing will be quicker, which helps to control the colours. Small-scale work on copper is generally based on gauge 22 – 0.71 mm (0.029 in). However, a thinner gauge can be used for some of the techniques, for example when the grounding is applied by dipping. Larger pieces require a base of gauge 20 – 0.91 mm (0.036 in) or gauge 18 – 1.22 mm (0.048 in).

Shaping a plaque for traditional Limoges-style painting

Limoges-style painted work is based on plaques which are gently domed or on suitably rounded shaped articles. On plaques the curve should be slightly more pronounced towards the centre, flattening towards the edges or rims. A narrow rim is turned down at right angles to the top surface to give strength to the plaque. (Occasionally the rim is turned up to contain the enamelwork.) Doming the plaque prevents warping and gives the finished piece the appearance of greater visual depth. The colours fire with a smoother, less undulating contour on the domed surface.

A plaque is prepared by cutting sheet metal to shape with a jeweller's saw, the corners being slightly rounded with a file. The plaque is annealed prior to shaping, then worked over a stake or anvil, using a wooden or leather (rawhide) mallet. The annealed plaque, if of a thin gauge, can be domed by placing it on a sand cushion or other resilient surface, such as a pad of newspapers, and rubbing firmly with a half-round burnisher of polished steel, working from the centre outwards. The edges are worked with the round end of a chasing hammer, to give a smooth turn with rounded corners. A proportion of traditional Limoges painted work was

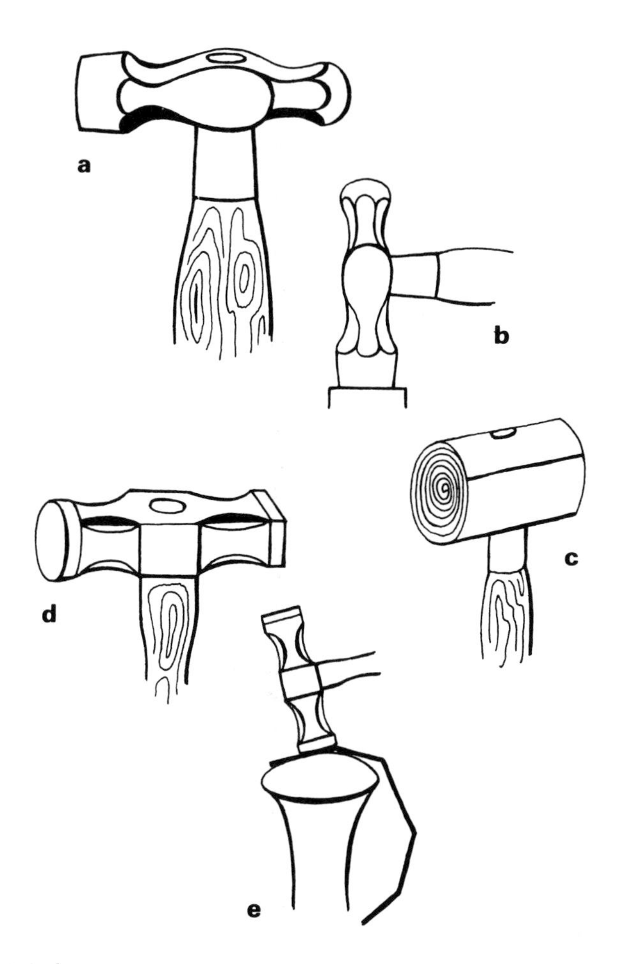

a *Ball pein hammer*
b *Correct position of hammer head when striking metal*
c *Rawhide mallet*
d *Planishing hammer*
e *Shaping metal over a stake*

based on straight-sided plaques with rims with soldered corners: for this the reverse side of the plaque is scored with a blunt chisel to ensure a good bend for the turned over rim. The squares formed at the corners of the plaque are cut away with a piercing saw, the rims are turned over a suitable block of wood or iron and the corners are soldered with hard silver solder.

For Limoges painted enamels the picture is built up with several fine layers of coloured enamels without physical barriers to contain them. However finely the pastes have been ground and applied, there will be

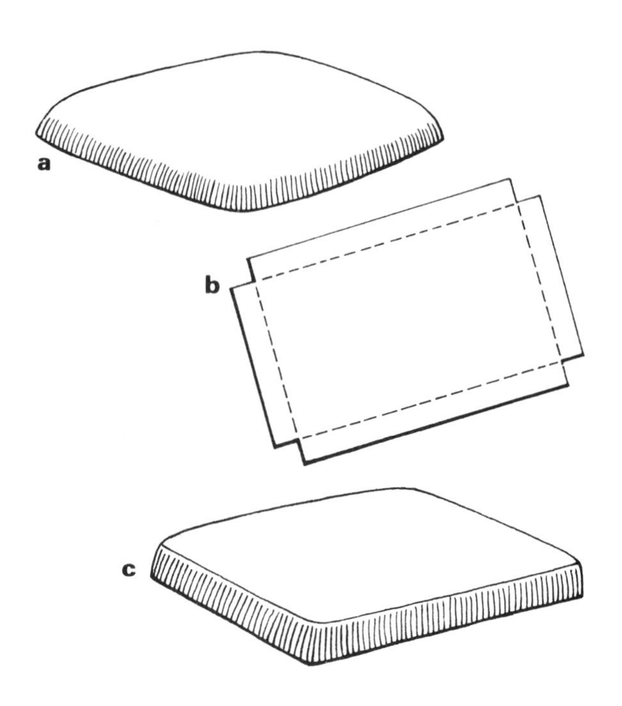

Limoges painted enamels
a *Plaque gently domed with turned down rims with rounded angles.*
b *Corners cut away when preparing a straight-sided plate for Limoges painted enamelwork*
c *Rims turned down and soldered to increase rigidity of the plate*

some degree of flow during the firing and it is difficult to produce precise outlines and sharp details. There are various devices which can be used to help give good definition. The traditional methods varied as each workshop had special techniques. One way was to scribe or scratch the picture into the clean copper base before covering the whole metal surface with flux. Where these guidelines showed through the flux grounding a dark enamel was laid in to form the outlines and the coloured pastes were filled into the spaces formed by the outlines. Finer details were added at a later stage with lines of gold. Greater freedom of design was possible by using a grounding of opaque enamel and applying coloured pastes over this. Before each layer was fired outlines could be incised with a pointed tool or needle. By using the needle it was possible to produce half-tones and shading by rubbing away sufficient of the unfired paste to allow the underglaze to show through. It became increasingly important to choose the firing sequence with care, so that the harder firing colours were fixed in the early stages and the design was completed with softer colours which

Painted enamel on a copper base. (Limoges, sixteenth century.) Courtesy of the Trustees of the British Museum.

were superimposed. In this way the layers did not sink down into each other on refiring. This method could produce a low relief effect in the surface of the enamelwork as highlights were built up with additional layers of paste. There were several variations of this method, the great master-craftsmen developing their own styles. The monochrome or grisaille enamels were produced by similar methods, with a light colour contrasted against a dark background, generally white against black.

Monochrome (grisaille) plaque in blue and white enamel, with overglazes and gilding. (Limoges, about 1650.)
Courtesy of the Victoria and Albert Museum.

Transparent effects were possible, despite the fact that the painted works were based on an opaque grounding, by using shaped pieces of gold or silver foil as underlays. The foils were positioned on the grounding, fired, then covered with transparent enamel and when refired this produced glittering patches of colour. Metallic oxides were at first applied sparingly as overglazes to the painted work, to tone or shade parts of the design and to add sharp outlines. Eventually the metallic oxides were used in their own right.

To make a Limoges-style painted enamel

The metal plaque is scoured with steel wool, pumice powder and water to clean it and give a slight key to the surface. After rinsing and drying the plaque is placed very briefly into the hot kiln to tarnish the surface slightly, which will degrease the surface. If the grounding is to be of flux the tarnishing is omitted and the metal surface is brightened.

The grounding of hard enamel is generally of flux or white. For monochrome paintings the grounding can be dark blue or black, the design being built up with layers of white. To apply the grounding a spatula can be used to spread the paste, to which two or three drops of gum have been added, leaving the metal evenly and smoothly covered. A quicker method is to apply dry powdered enamel over the gum coated surface, sieving the powder as evenly as possible (covering the edges and working inwards) and smoothing the layer down with the flat side of a spatula blade to press out trapped air. It is best to apply the grounding in two thin layers, separately fired. The fired surface may need smoothing with carborundum and refiring before the painted work is laid in. The counter-enamel should be of equal hardness and thickness.

If transparent enamels with highlights are to be included in the painted design then shaped pieces of gold or silver foil (paillons) are attached and fired to the grounding after the outlines have been indicated and before the colours are applied. The colours should be selected and test fired together to find which are the hardest and can be used in the earlier stages and how the colours will appear if superimposed over each other.

It is best to work from a drawing or tracing of the design made to the

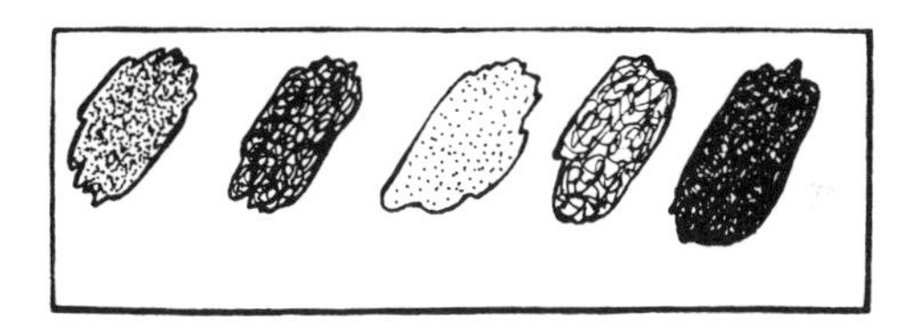

Painting enamels are test fired over a hard white enamel ground, to check their fusibility and compatability

correct scale. Outlines to show through the grounding can be traced and scratched into the metal with a scriber if flux is to be used as the grounding. For other types of grounding the outlines can be indicated on the glaze using a fusible 'ink'. This can be coloured metallic oxide mixed with gum, which gives a quick-drying mixture, or the oxide can be mixed with painting medium for better control. Such an ink can be applied with a fine brush or with a mapping pen, charging the nib with a brush. Finely powdered enamel mixed with gum or glycerine can be used in a similar way but the lines will not be as fine. There are ceramic pencils on the market which can be applied directly to the grounding and like the vitrifiable 'inks' they are fixed by firing. Another method is to prick along the outlines of the drawing which is then taped to the plaque and a small quantity of dry metallic oxide is forced (pounced) through the perforations using a flat brush, to leave a dotted line. When outlines are being laid in it is best to avoid dark lines where highlights will be needed.

Once the key points or the whole outlines of the design have been fixed the painting can be developed. The painted enamels are given greater naturalism and they can show more detail when the pigments are applied by brush. To make application by brush possible the finely powdered enamels are mixed with a suitable medium. A fairly heavy oil, such as the traditional fat oil, allows the pigment to be suspended evenly in the paste but such a paint is stiff to apply and dries off slowly. Another traditional medium, spike oil of lavender, is thinner and spreads more easily but sharp outlines are more difficult to control. These oils can be used together. Rectified spirits of turpentine can be added to the paint to thin it and speed up evaporation but too much thinner will leave a streaky colour. Modern painting mediums are offered and these include silk screen printing medium. The best colour and texture are obtained when the minimum of medium and thinner have been mixed into the paste. The medium is worked into the enamel with a spatula on a glass slab. All traces of the medium must have evaporated before the work is fired otherwise the colours will be filmed over due to fumes and blisters will appear where the oil bubbles up. If a thick paste is used, which flows off the brush with difficulty, the pigment can be applied in small spots and spread with a point. Sharper outlines are produced if a colour is allowed to dry before another is laid alongside. Oils should be allowed to dry off slowly with only a little warmth by placing the work under a lamp or near the kiln. The final stages of drying can be in front of the open muffle, letting the surface cool when fumes start to rise. When the work is warmed in front of the open muffle and no more fumes can be seen and the colours look pale and powdery the work is ready for firing.

When the coloured enamels are applied the layers should be kept as fine as possible. This will give the best colour and reduce the risk of cracks appearing. It is not necessary to cover the whole surface with colour for each firing. As extra layers are added to develop the design or give highlights a slight relief effect is produced. This can be slightly exaggerated to give a raised foreground against a flatter, darker background. Careful firing at every stage is necessary to keep the flow of the enamel to a minimum as otherwise the relief work flattens out and the outlines become less sharp. In the final stages details can be painted in with overglazes of metallic oxides or liquid gold.

The work should be well supported for firing in a cradle with only the rims in contact. Cooling should be slow.

The painted work is left fire-finished as filing would remove the upper layers of colour. The surface will be slightly undulating and very reflective.

Elaborate portrait miniature painted with metallic oxide overglazes on an enamel ground. (Portrait of Sir Sidney Godolphin, by Charles Boit 1663–1729.) Courtesy of the Trustees of the British Museum.

On-enamel or overglaze painting

There are many variations of the overglaze painting method. The final results may be achieved over many firings to produce the type of detailed miniature portraits or landscape pictures which were characteristic of the seventeenth, eighteenth and nineteenth centuries. Other effects are achieved with far fewer firings and generally give bright colour against a white or light background. This more spontaneous treatment has been used on a wide range of decorative articles from the seventeenth century onwards and continues in popularity. All types of box insets, jewellery items and small-scale shaped work can be produced. The more elaborate

Small-scale painting: overglazes on a grounding of enamel on copper. (10.2 cm × 12.7 cm (4 in × 5 in).) Dulwich College after Pissarro. By Susan Graus.

work is created with metallic oxide overglazes alone, but for work requiring only two or three firings the metallic oxides can be combined with finely ground coloured enamels to give greater brilliance or some translucency and texture to the design.

Small-scale painted work can be on flat metal, on gently domed plaques or on shaped articles. For elaborate work and pieces requiring many firings a gently domed plaque will be best as there will be no warping. There is no need to turn down a rim as for larger scaled, Limoges-type painted work. To dome a small plaque, the annealed metal is placed on a resilient surface and rubbed on the reverse side with a burnisher, working from the centre outwards. Straight-sided plaques should have the corners rounded off to reduce tension.

Plaques of gold or copper require counter-enamelling. The plaque is prepared for painting by firing on two thin layers of a high-fusing white enamel. Each layer is smoothed after firing, with carborundum stone. The second layer requires refiring after stoning to leave a smooth surface for painting.

The main outlines of the design can be laid in with freehand drawing, using a fine sable brush and prepared metallic oxides. Outlines should be omitted where pale colours and highlights will appear, for example clouds in a blue sky, as the darker lines tend to show through. To transfer outlines of a tracing taken from a same-scale drawing or photograph, the transfer

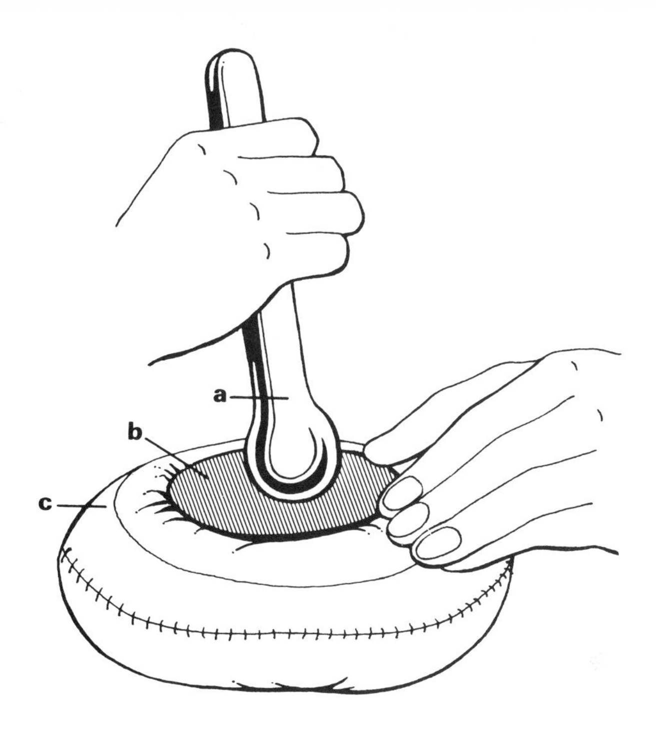

Doming a small plaque
a *Polished steel burnisher: rubbed from side to side until the centre of the plaque is raised into a gentle curve*
b *The annealed plaque*
c *Sand cushion*

Rectangular plaque for miniature painted work
The metal is gently domed and the corners are rounded. The plaque is prepared by applying and firing two fine layers of white enamel and counter-enamel.

Painting in outlines with metallic oxides

paper is pricked at key points, as described for Limoges-style painted enamels, and dry metallic oxides can be applied through the holes. Special pencils are available which are suitable for drawing directly onto the enamel grounding. Ordinary pencils should normally not be used as they may cause tiny blisters to form in the enamel during firing.

The prepared metallic oxides can be intermixed on the palette to produce a wide range of shades. However, there are some mixtures which do not give good results when fired due to chemical reactions and generally it is safer not to mix red or black with other colours. All combinations should be tested to see the fired tint which can be expected and to ensure that colours fired over each other will not adversely affect each other.

The painting medium has to be worked sparingly into the finely powdered metallic oxide pigments with care so that the colours are evenly distributed in the paste with no tiny lumps or streaks. Steel and iron may contaminate some of the metallic oxides and it is best to use stainless steel or glass tools for the preparation of the paints. Generally only a tiny amount of each colour is prepared – sufficient for the job in hand. For colours which will be used regularly a small stock can be prepared and this

The colours are developed over several firings, built up with thin layers of metallic oxides

can be stored in an air-tight bottle. A small quantity (to cover the spatula tip) of a colour is placed at the centre of the glass slab and the medium is placed at the perimeter, the medium being added gradually till the correct consistency is obtained. The spatula or glass muller is used to grind the powder and medium together, scraping the colour back into a little heap as it spreads out, and continuing in this way for two or more minutes, finally scraping the colour off the slab and placing it onto the glass plate or tile which acts as the palette. The grinding slab and spatula are cleaned with spirit and the next colour can be prepared in the same way. When setting the palette space should be left for colour mixing. The choice of medium depends on the method and type of painted work. For work which will be painted in slowly and with sharp outlines a thicker oil, such as fat oil, will give the best control. A light oil, such as spike of lavender is easier to paint with and evaporates more rapidly. When the basic colours have been prepared the intermediate shades can be mixed on the palette

Preparing metallic oxide overglaze on a glass slab for painting.

with a spatula or brush. If painted work is to be superimposed over a transparent ground it is necessary to mix the colour with white to produce a compact and dense paste and this will also, of course, give a lighter tint. To make a less dense pigment and to reduce the firing temperature needed to fix a colour it can be mixed with printing flux – this special quality of very finely ground enamel will also impart a degree of gloss to the otherwise matt metallic oxide pigments. Fine brushes are used to lay in the painted work. Sable brushes of size 000 upwards are suitable and they should be good quality brushes which form a fine point. Brushes are cleaned with methylated spirit after use, followed by a quick dip into a thin oil to keep them supple.

Provided dark colours are not painted in too early and more fusible pigments are not applied with too high a firing, it is possible to paint in a fairly developed design from the beginning. Some artists prefer to begin with an underpainting, applying the broader areas of colour, using the

Overglaze painting on a grounding of enamel on copper. Watches, c. 1710. By courtesy of the Ashmolean Museum.

higher firing pigments, following by shading, highlights and details in later firings.

For each firing the work needs to be completely dried, so that on checking by holding the work in front of the open muffle, no fumes can be seen rising from the surface. The dry colours will look very pale with a yellowish tinge and powdery. After firing many of the pigments will have lost some of their vigour or depth. To develop some of the richer hues it may be necessary to repaint and refire several times. A single thicker application of pigment would not produce the correct tones and would tend to blister or craze.

Firing of the miniature painted work is very quick. Metallic oxides fire in the range of 720–750°C (1328–1382°F). As these pigments do not become shiny with firing, care is needed not to risk overfiring. If underfired, the painted area looks powdery or grainy and requires a few further seconds in the muffle. Overfiring results in loss of colour and may allow the painted work to sink down into the grounding.

Because the fired metallic oxides are matt in appearance and because they are not as durable in use as true enamels, some types of miniature painted work are given a final glazing or crowning of a clear, soft flux. Practice and care is needed if this is attempted: if the flux is applied too thickly it will veil the painting beneath and if too coarse a powder is applied it may disperse the painted surface. The flux is traditionally applied in two very thin, separately fired layers. Each layer (after firing) is filed with carborundum stone under running water, to thin it down still further. The final gloss is produced by refiring or by polishing. Generally, however, small painted enamels are left without such a finishing glaze and they are not polished as this would remove some of the pigments.

When metallic oxides are used in conjunction with overglaze enamels the work is generally completed in two or three firings, working on the prefired grounding. The metallic oxides are painted on to give outlines and the enamels are laid in to give patches of glossy colour. For example, in a design of a floral spray, dark green and brown metallic oxides would be painted on to show the stalk and leaves and outlines of the petals. After firing, coloured transparent pink and green enamels, prepared as pastes, would be laid in to cover the flower heads and leaves, keeping within the outlines to allow for some flow during firing. After this firing, finishing touches such as the veins on the leaves can be painted in with oxides and the work is given its final short firing.

Underglaze pigments and pencils

Underglaze lead-free black pigment is offered by some suppliers in the form of a dry powder and requires mixing with oil or a water-based medium. The paint which this produces is suitable for shading and details which can be applied directly to a copper surface and then covered with flux.

Underglaze pencils are richly pigmented ceramic pencils which can be used to draw details and shading on a prefired enamel surface which should be slightly roughened by stoning.

Transfers

Lithographic transfers suitable for firing to a prefired enamel surface are supplied in the form of small pictures or motifs made from very fine layers of metallic oxides fixed to a temporary backing and covered with a protective film. The strongest effects are produced over a white enamel grounding, but other opaque enamels can be used for the grounding. Gold transfers are also available and these are effective on dark opaque grounds.

Manufacturers' recommendations should be followed to obtain good results with transfers. Generally a transfer is prepared by soaking in a saucer of tepid water until the backing paper curls. The transfer is then positioned on the grounding (paper side uppermost) and the design is deposited on the enamel by sliding off the paper. The design is pressed onto the grounding with a damp cloth pad, to squeeze out air pockets and draw off excess moisture. Slow drying, preferably for several hours, is needed. When dry a short firing at about 725°C (1337°F) is given, leaving the kiln door open for the first few seconds to allow fumes to escape as the holding film burns off. The transfer design can be used as an underpainting, the work being completed by overpainting with metallic oxides applied by brush.

MODERN METHODS AND EXPERIMENTAL IDEAS

Modern enamelwork on bases of gold and silver

Contemporary influences are echoed in enamelwork and present-day enamelled designs on gold and silver display inlays of clear, lustrous colour which follow the main outlines of the metalwork, so that the enamel is an integral part of the article. There are changes in the choice and distribution of colour compared to work of the past and designs tend to be broader. The actual methods continue to be traditional, mainly in basse taille, champlevé and cloisonné. These methods allow the display of high-quality craftsmanship and ensure durable and finely finished work suitable for costly pieces. Silver is generally used, due to the very high cost of gold, and it can be gilded. Transparent qualities of enamel are the most popular as they allow the precious metal to show through. It is, of course, a great advantage that when enamel is inlaid into the metal there is a saving of the gold or silver required for the finished article, yet the work will feel weighty and more valuable due to the enamel.

Experimental ideas

The experimental work is based on copper and sometimes on steel. Small-scale work is very wide ranging, the designs including brilliant colour applied over etched metal, transfer patterns and small painted pictures. Large-scale work can be particularly exciting to make and can produce effects not obtainable with other materials. Unique pieces can be produced by using enamel as an artistic medium, applying the colours to the metal according to a general plan but working out the final colour balance and textural effects by adding colours or manipulating the enamels stage by stage. This type of design is based on a grounding of enamel which can be quickly applied by dusting on dry enamel or by using a slurry sprayed over the surface with an airbrush.

Pendants: Shaped, etched and soldered copper bases, covered with opaque and transparent enamels, partly over silver foil.

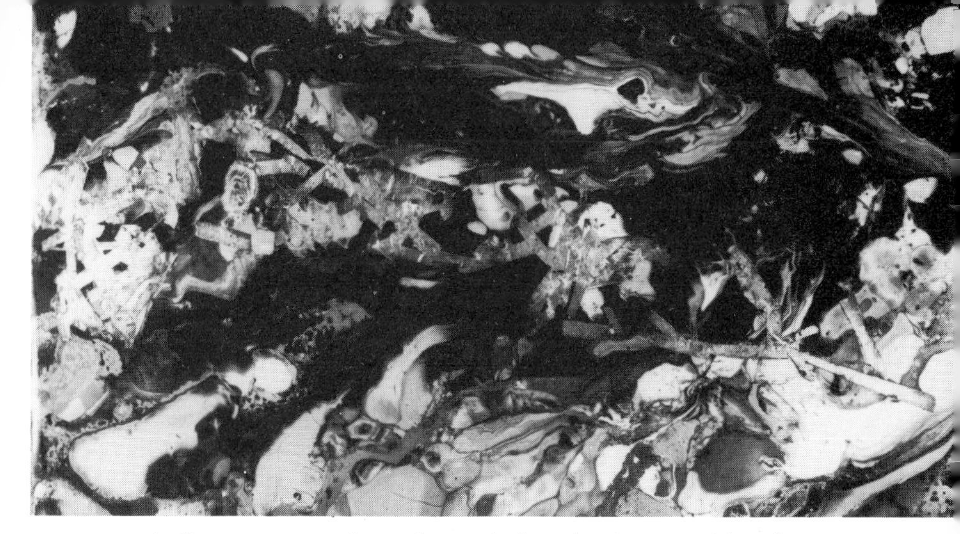

Experimental effects: various qualities of enamel allowed to interact with each other during high firing. Flow-patterns in shades of blue with silver foil. By Erika Speel.

Covering the metal surface by spraying

If an airbrush is to be used, it is necessary to have a spraying booth and to work in a well-ventilated area, away from heat or open flames as a volatile, flamable mixture will be applied. Very finely powdered enamel can be sprayed providing the mixture is fine enough to pass through the needle of the air brush. Metallic oxides can be used in a similar way, for spraying over a prefired enamel grounding. The pigment is mixed into a paint-like consistency with water and alcohol: 50 parts of finely powdered pigment to 35 parts water and 15 parts of denatured alcohol. The nozzle of the air-brush is moved about to spray the paint evenly and deposit a fine coating of colour, holding the nozzle about 15 cm (6 in) from the surface of the plaque. The depth of colour is built up by spraying over the same area repeatedly, to deposit several thin layers. For line effects a very fine spray is needed while a wider spray gives surface cover. The work is dried and fired in the usual manner.

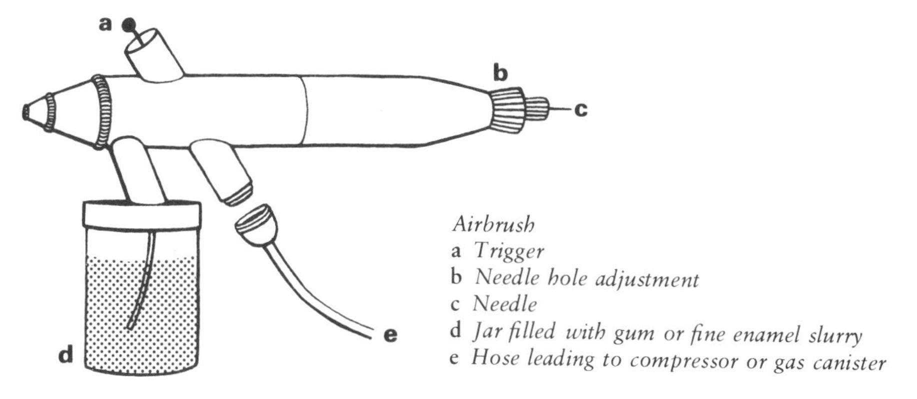

Airbrush
a *Trigger*
b *Needle hole adjustment*
c *Needle*
d *Jar filled with gum or fine enamel slurry*
e *Hose leading to compressor or gas canister*

Enamelling flat copper plaques

Gauge 18 – 1.22 mm (0.048 in) is suitable for flat sheet copper which is to be covered with enamel. In this thickness larger format work is not difficult to handle and the metal can heat up quickly which is an advantage in controlling the coloured enamels. A slightly heavier gauge is suitable for work which is to be only partly covered with enamel and where the design will be engraved or etched into the surface. The heavier gauge metal will be less subject to warping and counter-enamelling can be omitted, but slightly longer firing will be needed at every stage.

To control the copper surface it is necessary to apply counter-enamel of approximately equal thickness to the top layers of enamel when copper gauge of 18 or thinner is used. It is also generally necessary to flatten such work after firing by placing suitable weights on the surface. To flatten the metal, the plaque is lifted onto the cooling slab with a long spatula and as soon as a skin forms on the enamel (this happens very quickly when there is only a thin layer of enamel) but before the metal has lost too much heat, the weight is pressed onto the surface and left in position until the work cools. When there is a greater thickness of enamel at the centre of the plaque, the corners tend to curve upwards and these can be flattened by using two weights at the edges.

There is some tendency for the edges of flat plaques to become ragged as the enamel shrinks inwards during firing, leaving a darkish rim. To avoid shrinkage the metal should be well scoured before enamel is applied and after each firing the edges should be inspected and more enamel added if thin patches are visible. If a dark rim builds up, the plaque should be pickled and if necessary the discoloured ridge can be removed with carborundum and water.

Designs for copper can be by any of the methods or combinations of methods adapted for the larger scale. Wire outlines can zone the coloured

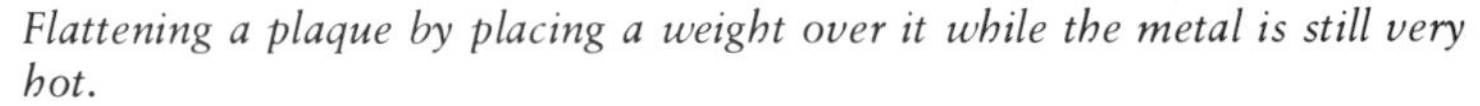
Flattening a plaque by placing a weight over it while the metal is still very hot.

areas and broader designs can be created with sgraffito, scrolling or collage work. Where possible the thickness of the enamel should be evenly distributed over the surface to prevent warping and the formation of cracks due to tension.

Enamelling on a base of steel

Steel of a special mild quality can be used for enamels. Steel is very suitable for larger format work and for architectural enamels such as murals or door plates, which may be sited outdoors. Steel of gauge 20 – 0.91 mm (0.036 in) is used for general purposes.

Steel panels and discs, prepared with a grip-coat and white enamel grounding can be bought and these require only the addition of the decorative work. If the steel base is to be prepared in the workshop some extra work is needed. To ensure a good bond between the metal and the enamel the steel surface requires degreasing, roughening and firing with a grip-coat before the grounding and subsequent decorative work are applied. The steel is degreased by heating for three to four minutes in the muffle at about 700°C (1292°F). To roughen the surface the cooled metal is immersed in a 10% hydrochloric acid pickle until it is a light grey colour, and, after rinsing off, it requires neutralizing by dipping into a bath of 1% solution of sodium carbonate, followed by rinsing in hot water and drying. This leaves the steel ready for application of the grip-coat. The grip-coat is a special quality of enamel suppled either as a dry powder which has to be mixed into a slurry with water, or as a ready mixed liquid. The easiest way of applying the grip-coat is to dip the steel plate or article into the slurry, using a shallow dish, and then to stand the plate in a vertical position to let it drain. The grip-coat will cover the top and counter sides of the metal and no further counter-enamelling will be necessary. When the grip-coat is dry the plate can be fired on a stilt support at about 840°C (1544°F). When cool, the grounding of opaque white enamel is dusted or sprayed over the top surface in a thin layer, dried and fired at normal enamelling temperatures. The decoration is then applied in successive fine layers, in two or three firings.

Ordinary jewellery enamels are suitable for use on a base of steel prepared in this way but transparent effects will be very restricted over the opaque base. There are also special opaque steel enamels which can fire in the range 800–820°C (1436–1472°F).

Designs can be applied by stencilling, sgraffito or transfers, with colours applied by dry dusting or by air-brush techniques.

Broad effect cloisonné

A development of the true cloisonné method is possible by composing designs in which only the main outlines or special features are created in wirework or cloisons but without enclosing each separate colour. Extra outlines can be added at different stages of completing the design and this gives artistic freedom. It is not necessary to enclose each separate colour in

Broad effect cloisonné. (St Francis, by Hilde Hamann.)

compartments and, indeed, some flow of colours where they meet may be encouraged. The cloisons can be left slightly higher than the enamel inlay, to add an extra textural feature or to cast shadows. The surface can be left filled to different levels, so that the enclosed areas may be a little higher than the surrounding field. These methods leave a brilliant, undulating surface and are best suited to larger scaled work.

Metal frets

A fired enamel surface can be partially overlaid with a metal design formed from shaped wires or pierced out sheet metal or foil. Sheet metal can only be successfully embedded if used in the form of thin strips or very thin gauge metal, otherwise it tends either to sink down too deeply or warp and break off in cooling. A good counter-enamel is required to prevent the metal base from warping and flat work should be weighted down during cooling.

The metal fret is attached to the fired enamel surface and the piece is returned to the muffle. When the enamel re-melts, a long-bladed metal

a *Design with embedded gold and silver foil with polychrome enamel.*
b *Design with embedded heavy gauge flat wire against a dark blue ground.*

spatula is used to gently press down on the fret to help it sink sufficiently deeply into the surface of the enamel to ensure adhesion. If the fret has been cut from gold or silver foil it should be allowed to attach to the surface of the enamel without any such pressure and if on cooling it appears that the foil has not attached everywhere the free portion can be burnished to lie flat on the enamel and the work then requires refiring. Care is always needed when using silver foil as the edges may melt if too highly fired. Superimposed frets of copper or silver can be lightly varnished after completion of the piece, to prevent darkening of the metal surface.

The finished appearance of the work will resemble filigree if round wires have been used, or inlaid work if flat wires or foils have been embedded.

Free-form or unstructured designs with enamels

A copper surface prepared with a grounding of hard white or, more frequently, with a hard flux, can be decorated with unstructured designs by allowing the coloured enamels to interact with each other in various ways. Finely pulverized colours can be dusted on in successive stages to build up into shaded or contrasting 'washes' of colour. Speckled or mosaic-like effects can be produced by applying small pieces of variously coloured enamel fairly closely together – held in position with a little gum if necessary – and allowing these to spread out into a smooth surface with high firing. If the resulting effects of colour need 'pulling together' to increase the visual interest or to give balance, outlines or highlights can be added with metallic oxide overglazes or liquid gold.

Stencils

Stencils are particularly useful if a pattern has to be repeated on groups of articles. A stencil is any suitable shape cut from paper, card or metal, which is laid over a prefired enamel ground and used as a mask so that when a layer of a contrasting colour is applied and the stencil is lifted off, a silhouette design will be left in the original grounding.

Stencils cut from absorbent paper, such as newspaper, give the sharpest outlines and can be moulded to follow a shaped surface. Small stencils may be difficult to lift off unless tabs of sticky paper are attached to their tops. For flat work, machined shapes such as metal cog wheels from watches, or natural shapes such as leaves, can make interesting stencils. Netting, open weave cloth, pieces of string or threads arranged in loops over the grounding offer other variations.

Before applying a stencil the fired grounding is degreased if necessary and painted over with gum. The stencil is dipped into gum and placed on the grounding. If parts of the grounding are to be left in the original colour, these can be masked off with pieces of paper. Before the gum dries a fine layer of dry enamel is sieved over the surface and before this layer is quite dry the stencil is lifted off vertically. When dry the work can be fired. A short firing will leave a sharp outline which softens with a longer firing. Crispness of outline will also be affected by the hardness of the enamel colours which are being used. The design can be developed by re-applying the stencil, off-setting it against the first silhouette, using toning or contrasting colours. Variations can be produced by applying the negative stencil – the piece of paper from which the design was cut out, or by cutting the stencil into sections and using the parts to echo the main motif.

Sgraffito

This method is similar to the treatment of some types of glazed ceramic work from which this technique takes its name. A prefired, gum-coated enamel grounding is covered with a layer of finely powdered dry enamel in a contrasting colour. Before this layer dries, a line design is scratched into it with a pointed tool to reveal the glaze beneath. The loosened powder is gently blown or brushed away and stray specks of enamel powder can be lifted off with a moistened paint brush. The lines which are incised should be drawn in slowly with the tool to prevent the outlines becoming ragged. Lines which are very fine tend to merge together in the firing unless a very thin layer of hard enamel has been dusted on. A short firing is given to fix the design and the scratched in lines will be slightly recessed.

Collage or assembled designs

Lumps of enamel, glass beads, mosaic tesserae, enamel threads and other glass decorations can be applied and fired to embed in an enamel grounding. If the decorative material is made of a harder glass, such as pieces of stained glass, permanent adherence is difficult. If overloaded, designs may be too richly coloured, shiny or bumpy.

Millefiore beads, enamel threads and lump enamel embed most readily into the grounding, smoothing to become flatter with prolonged firing. Specially made glassy inclusions can be bought from craft suppliers. These are offered in thin squares or rectangles or as small spheres in transparent and opaque colours. Transparent inclusions can give a jewelled effect if fired over a flux ground which has been covered with a soft white enamel. When the inclusions are fired to embed on this surface, they sink down through the white layer and being based over the flux and surrounded by white the brilliance of the transparents is increased. Millefiore beads are bought as assorted pieces, varying in size, thickness and pattern. They are composed of concentric bands of easily fusible glass. A tightly organized pattern is produced when millefiore beads are given a short firing and they can spread out to form softer patterns with prolonged firing. If set closely together the millefiore beads can be filed after firing to give a smooth surface.

Enamel threads, glass tesserae, millefiore beads and two fired beads over a grounding of flux on copper.

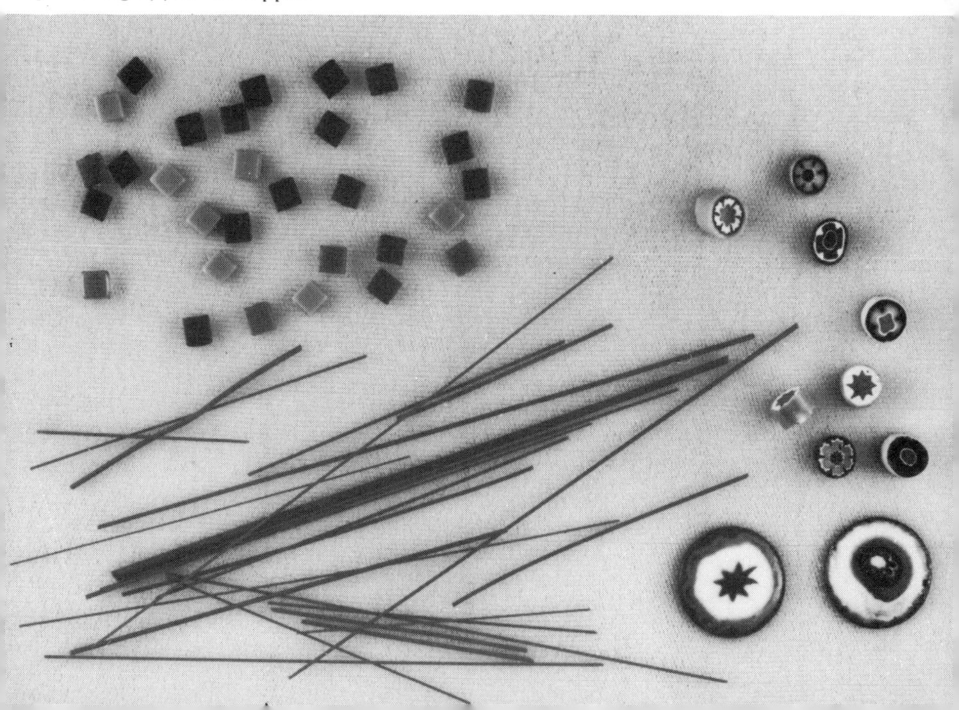

Lump enamel

Lump enamel can be fired over a pre-enamelled surface and will fire above a hard grounding or sink down into a softer grounding. Lumps larger than about one millimetre (0.05 in) tend to split into smaller fragments unless fired slowly. The pools of thick enamel produced by firing on larger lumps tend to crack after cooling. Small lumps of transparent enamels fired over foil can produce the appearance of cabochon (rounded) jewels.

Scrolling

While a layer of enamel is hot enough to leave it in a molten state, it can be drawn out in scrolls or swirls to merge with surrounding colours or to leave spiked, spiralling or looping trails. The usual method is to fire some lump enamel or small heaps of powdered enamel on a prefired grounding, arranging these where the strongest colour will be required and inserting the work into the muffle to allow the surface to melt: where the pools of enamel form, the point of the steel scrolling tool is inserted, but not deeply enough to touch the metal base, and by moving the point in a circular, figure-of-eight or feathering gesture, the molten colour is worked into scrolls, swirls or flow patterns. If the scrolling tool is long and the kiln has enclosed elements or is of the type where the power switches off when the door is open, the scrolling can be done while the plaque is still in the muffle. Otherwise the scrolling can be done immediately the work has been withdrawn and while the enamel is still sufficiently soft. The piece can then be briefly refired to let the surface smooth out.

Pendant decorated with a scrolled design in opaque colours over a grounding of translucent green enamel.

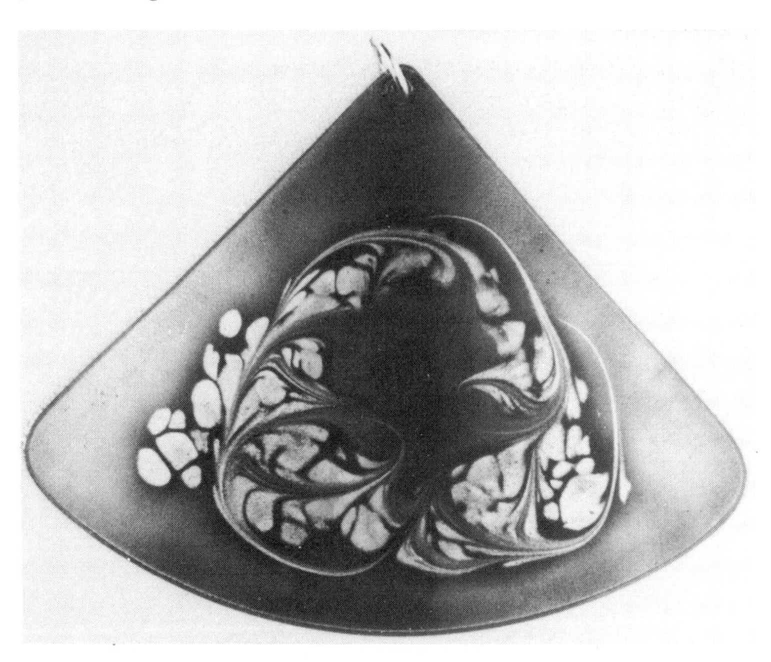

Enamel threads

Enamel threads can be bought from craft shops and are supplied in small packets containing a single colour or a mixture, in varying thicknesses. Enamel threads can be made in the workshop by melting lump enamel on a piece of scrap copper and using the pointed scrolling tool to draw out small globules of enamel – the longer the pull the thinner the thread which forms will be. The thread hardens immediately and is broken off with tweezers. The point of the tool is cleaned by quenching. Enamel threads can be heated and bent to shape with tweezers.

PRECIOUS METAL FOILS AND LIQUID METALS (LUSTRES)

Foils

Pure gold and silver is obtainable in the form of very thin sheets or foils. These can be cut to shape and used in conjunction with transparent enamels to increase the range of these colours. By using the foils highlights or more intense colours can be given where required and the full brilliance of transparent enamels can be shown even when working on an opaque or non-reflective ground.

The foils are supplied in small sheets packed between protective layers of paper. The cut-out shapes, or paillons, are prepared by drawing the outline on paper then sandwiching the foil between the cover paper and a backing paper and cutting out the shape with a sharp scalpel or scissors.

Gold and silver foil sheets are prepared and cut to shape while sandwiched between protective cover papers.

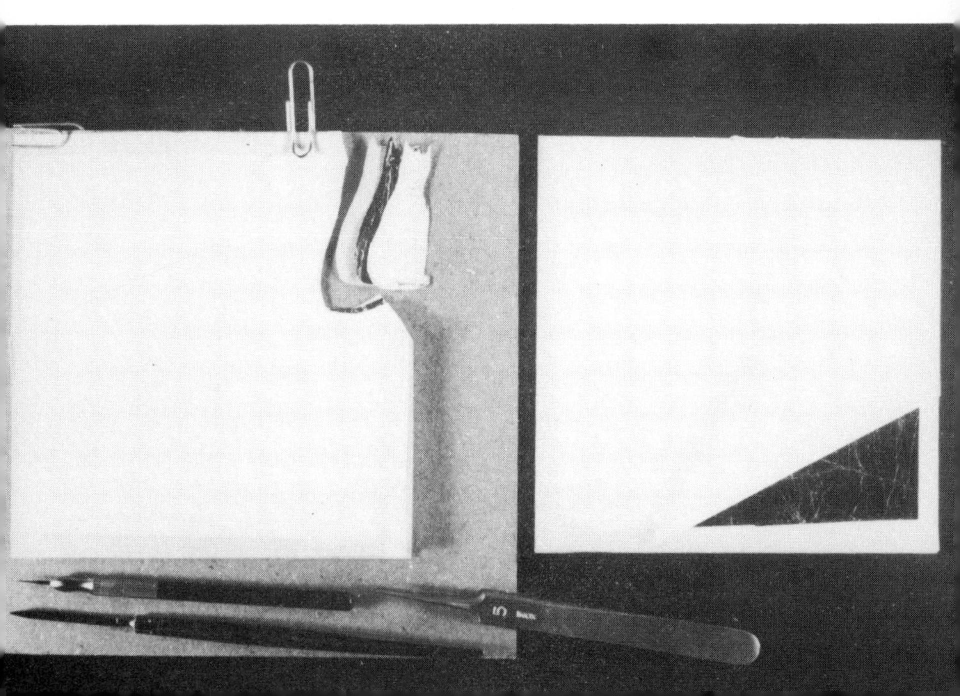

The foils are never touched by hand and are prepared and handled between the cover papers, and lifted with tweezers. Larger paillons have to be pricked over their whole surfaces with a fine needle point to leave vent holes for the escape of gases during firing, otherwise blisters will develop. Several needles can be inserted into a cork, their points protruding to an even level, to make the pricking quicker. Foils can be annealed and moulded to fit shaped articles. To increase their reflective properties the foils can be slightly crinkled and for smooth effects the foils can be burnished after being fixed by firing. Foils are not applied directly to the metal base but are positioned on the prefired enamel grounding, held in position with a little gum at the edges and fixed with a brief firing.

Firing of silver foil should be very quick to prevent the thin metal melting. After firing, areas of silver foil can be scoured with fine steel wool to roughen the surface and give a better hold to the superimposed enamel. When covering the foils, well-washed, finely powdered coloured enamels or flux should be applied in two thin, separately fired layers for greatest transparency.

Shaped pieces of foil can be embedded into the final layer of enamel without covering with a glaze for some of the modern methods.

Liquid metals or lustres

The lustres are obtainable in the form of a liquid which contains the very finely pulverized, pure metals suspended in a painting medium or vehicle. A few of the lustres are now being offered in the form of a paste. Liquid gold may be offered in various qualities, the fired finish ranging from a high gloss to matt and some types are suitable for burnishing.

Lustres show to greatest advantage over opaque enamel, especially when very fine lines are applied. Gold lustre is generally the most effective

Liquid precious metal and thinning solution

of these overglazes. Outlines can be drawn in with the lustre or shading effects or surface cover can be given.

The mediums in which the lustres are suspended are volatile and if the mixture thickens a small proportion of an appropriate thinning solution can be added. Excessive thinning should be avoided. The phial containing the lustre should be shaken and the liquid is then applied as an overglaze straight from the container, using a sable painting brush or mapping pen. The lustre-painted surface should be completely dry before firing. The kiln door should be left slightly ajar for the first few seconds to allow the escape of fumes.

Manufacturers' recommendations on firing should be followed as different qualities may require specific firing temperatures. Usually about 800°C (1472°F) will be required to fix the lustres in a fairly quick firing. If underfired the lustres will not adhere permanently to the enamel surface or they may appear dull. Overfiring tends to burn out the lustres so that another layer has to be applied. Slight overfiring gives subdued brilliance, especially with gold lustre, and this can be useful in producing a mellow tone. If too thinly applied the lustres will not show effectively. If too thickly applied there is less control of the design and before evaporating the liquid may spread out. When applied over a thick layer of enamel the lustre tends to craze during the refiring due to expansion of the grounding.

Liquid gold and liquid silver (lustres) applied as overglazes to reinforce outlines and add highlights to the design.

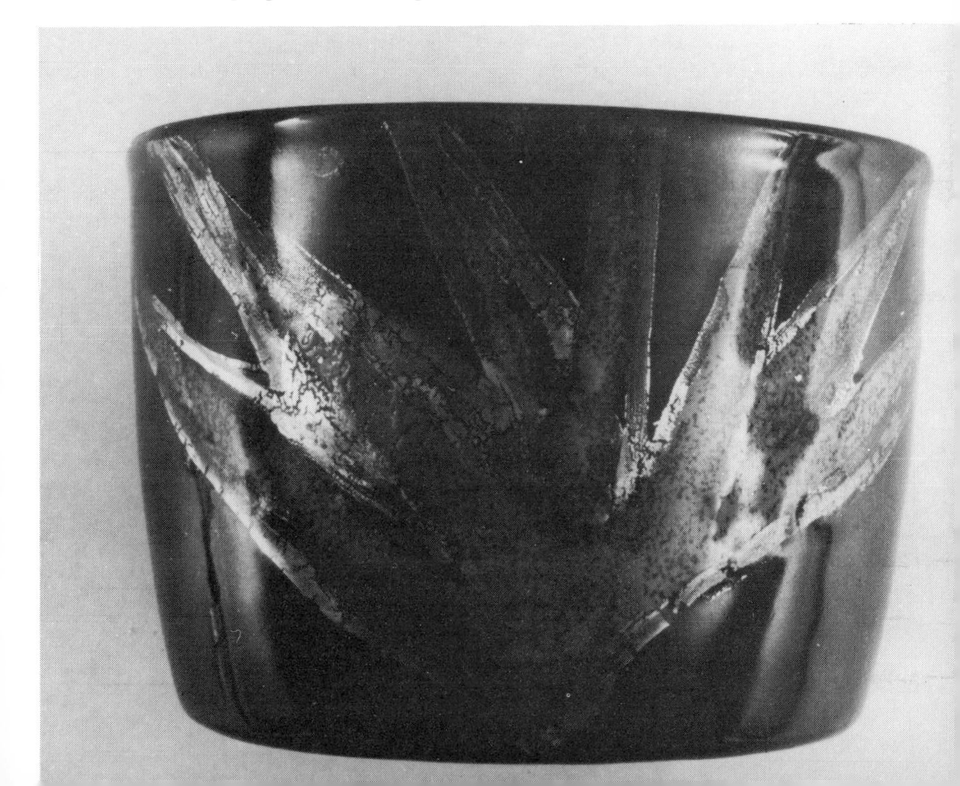

4 Projects

ENAMELLING A FLAT COPPER PENDANT WITH A STENCIL DESIGN

1 Choose a copper pendant of gauge 18 with a suitable hanging hole. Smooth the rims with emery paper if necessary. Scour the top and reverse (counter) side with fine steel wool and pumice powder. Rinse well and dry the metal.
2 Place the pendant on a firing support and place it briefly into the hot muffle, removing it as soon as the metal begins to colour. This will ensure a grease-free surface. Leave the pendant on the support to cool and subsequently handle it with tweezers.
3 A W-shaped or stilt-type support will be needed with arms angled to hold the pendant by its rims. A second tray-type support will be required on which the stilt can stand.
4 Dry powdered enamels will be needed in two contrasting colours, such as a high-fusing dark blue for the grounding and counter-enamel and a softer fusing white for the top layer.
5 Cut a stencil shape of suitable proportions and with simple outlines from soft paper. Attach a lifting tab to the top with sticky tape to make removal easy.
6 Brush the top surface of the pendant with gum. Sift blue (or other shade) enamel evenly over the metal, covering the edges first and working towards the centre.
7 Check that the hanging hole has not become clogged and remove stray specks of enamel from the metal rims. Place the pendant on the flat support and leave near the kiln to dry and warm.
8 When ready a brief firing is given, the piece being removed from the kiln as soon as the enamel begins to flow into a smooth glaze. The pendant is removed from the support and left to cool on a metal slab with a weight over it to prevent warping.
9 When cool, the fire-scale is removed from the bare metal areas of the pendant. The counter-side is covered with blue enamel in the same way as for the top grounding, but giving slightly thicker cover, especially at the edges. The pendant is placed counter-side facing upwards onto the stilt support. When dry the piece is refired as briefly as possible.
10 When cool the rims of the pendant are cleaned.
11 The top surface of the pendant is brushed with gum. The stencil is dipped into gum and positioned on the blue ground. White enamel is

Lifting off the paper stencil.

dusted over the surface up to and slightly over the edge of the stencil. After a few moments, before the gum has dried completely, the stencil is lifted off with tweezers, leaving a blue silhouette design where the stencil had masked the surface. Any stray specks of white enamel are removed with a moist brush.

12 When quite dry and well warmed the pendant is refired on the stilt support, top surface upwards. A slightly stronger firing can be given but the pendant must be removed from the muffle before the sharpness of the outlines softens.

13 Finish by rubbing the metal rims with emery paper and cleaning the hanging hole with a round needle file.

ENAMELLING A DEEP BOWL WITH A FREE-FORM PATTERN

1 Choose a copper bowl of gauge 20 – 0.91 mm (0.036 in). The top rim should be level so that the bowl can be fired either way up.

2 Prepare a mesh tray-type support on which the bowl will be able to stand the right way up (base downwards) or inverted (base uppermost). Check that the joint height of the bowl and support will leave good clearance when inserted into the muffle.

3 Dry, powdered enamels will be used. The grounding will be a high-fusing flux, the main colour will be transparent dark blue, the contrast will be given with opaque red and purple, the highlights will be added with fine lines of opaque white or gold lustre.

4 Two paper masks should be cut to fit the inside and outside dimensions of the base. The mask for the inside of the bowl should have two small tabs attached and it is used to catch excess powder while the inner sides of the bowl are being dusted over. The mask for the outside of the base keeps the metal free of enamel in the early stages so that the bowl can stand either way up during firing.

5 The bowl is cleaned and brightened by scouring or with acid. After cleaning the surface should be rubbed with steel wool to roughen it slightly. This is followed by rinsing and drying.

6 Brush the inside of the bowl with gum working upwards towards the rim and place the paper mask over the base.

7 Working quickly before the gum dries, sieve flux evenly over the interior of the bowl, holding the bowl by its base and rim, tilting and rotating it so that the flux is evenly dusted over the inner surface. Gently remove specks of enamel from the rim. Lift out the masking paper, removing any enamel powder on it. Then lightly cover the bottom of the bowl.

8 Place the bowl, base downwards, on the firing support. When quite dry and well warmed insert the piece into the muffle. Give a brief firing, just sufficient to let the enamel flow, then remove the work from the muffle and let it cool slowly on the support. (Taking the bowl off its support too soon may lead to a distortion of the shape.)

9 When cool, the outside of the bowl is cleaned and brightened. The masking paper is placed over the base of the inverted bowl and the sides

are covered with flux. If the bowl stands on a wooden block or on the blade of a spatula it is easier to tilt it while the enamel is being dusted over.
10 When the gum has dried the bowl is lifted onto the support, base uppermost. A short firing is given. The quick firing may leave the flux a little cloudy or there may be slight undulations in the glaze, but subsequent firing will improve this. If there are any unduly thick patches of enamel they should be smoothed by stoning with carborundum and water.
11 The colours are then applied in stages. The inside and the outside of the bowl can be covered for the same firing, or this can be done in alternate firings. The bowl should stand right way up and inverted for alternate firings to prevent the enamel flowing in the same direction and thickening at the rims. To produce the free-form effect, paint medium thick gum over the areas where the colour is to adhere, in loops, parallel lines or bands, without definite edging. Before the gum dries dust transparent blue over the tacky surface so that the powder adheres. If powder remains where not required it is brushed or blown off. When dry the design is fixed by firing.
12 Next, apply a small proportion of red enamel – it is usually best to apply opaque red very sparingly, adding more in an extra firing if necessary. The red is applied to run roughly parallel to the blue glaze. When this has been fixed, purple is applied to complete the main areas of the design – this colour can provide a counter-point or it can be used to merge harsh contrasts produced by the other two colours against the flux background. The purple is fixed by firing, this time the firing can be a little sharper to produce a smooth glaze.

Completing the design with white highlights on a deep bowl.

13 When the balance of the design is satisfactory the bare (exterior) metal base is covered with enamel. First the metal is scoured to leave the surface clean and roughened, next brush the bare area with gum and dust dark blue enamel over this ensuring that the powder overlaps a little with the fired surface so that no join will be visible after refiring. Fire the bowl with the base uppermost from now on. After cooling, apply a second thin layer of blue over the base and refire.

14 Highlights are added to the design to superimpose a strong linear pattern or to give some separation of the coloured areas. This can be done by applying streaks of medium thick gum to the glazed surface, using a fine brush, then dusting finely powdered white enamel over. Excess powder is carefully brushed off with a soft dry brush or a feather. The piece is refired when dry. If liquid gold lustre is to be used for the highlights, this is painted on to the fired glaze with a finely pointed brush (the phial of lustre will require good shaking before use). The lustre is left to dry slowly and it is fixed by firing at about 800°C (1472°F) – leaving the kiln door ajar for the first few seconds and ventilating the workroom to ensure correct firing and to allow fumes to escape.

15 After the last firing the metal rim of the bowl is brightened by rubbing with emery paper.

DECORATING A LARGE COPPER PLAQUE WITH A CLOISONNÉ DESIGN

1 Choose a rectangular copper plate, of gauge 18 – 1.22 mm (0.048 in) with a scratch-free surface.

2 A cradle-type support will be needed.

3 Draw the outlines of the plate onto a sheet of cardboard, then transfer the design onto this, and cover with double-sided transparent sticky tape. As the cloisons are bent to shape they are held in their correct position on the tacky surface until needed.

4 Anneal, clean and shape the cloisons, preferably using flattened copper wire of gauge 20 – 0.91 mm (0.036 in). Leave at least one curve or bend for each separate piece, so that it will stand easily when being fired to embed.

5 Clean and brighten the top surface of the plate. Coat the top of the plate with gum and dust over the whole surface with a smooth layer of high-firing, dry, powdered flux, giving a little thicker cover at the edges of the plate.

6 Place the plate on the support and warm it well. Let the muffle heat to about 950°C (1742°F) to allow for heat loss when the piece is inserted, otherwise firing will be unduly slow. A large piece will require turning midway during the firing to ensure even firing. The piece, on its support, is withdrawn to be able to do this and given a half-turn before re-insertion to complete the firing. It is necessary to complete this operation quickly before fire-scale can form on the exposed copper areas, otherwise the piece may have to be cooled and cleaned before being returned to the muffle.

7 When fired, remove the piece from its support and position a weight over it before it cools too much.
8 After cooling and cleaning of the bare metal areas, a second fine layer of flux is applied and fired to the top surface.
9 After cooling and cleaning, the counter-enamel is applied, using a high-fusing enamel. The depth of the counter-enamel should be roughly equal to the thickness of the flux on the top surface. If the design on the top requires many closely set cloisons, then pieces of copper wire can be arranged on the counter-enamel at this stage (i.e. counter-cloisons) which will embed during this firing and will help to prevent warping.
10 To fuse the counter-enamel, the plate is placed on a U-shaped support, counter-side upwards, the rims of the plate only in contact. The firing should only be long enough to give an orange-peel surface to the counter-enamel as it will smooth out subsequently. If counter-cloisons have been embedded the plate will require weighting down during cooling.
11 The plate should now be ready for attaching the cloisons, creating the outlines of the design. The flux grounding may require filing with carborundum, to ensure that the cloisons can stand on a level surface. The grounding is then brushed with a medium thick gum. Using tweezers, each cloison is placed in position on the grounding. When all the cloisons for that firing are in position a fine layer of flux is sieved over the surface. When quite dry, stray grains of flux are brushed off the tops of the cloisons.
12 The firing which follows allows the cloisons to embed into the flux. From now on the plate has to be fired on a stilt or U-shaped support and positioned right way up only. The plate requires warming before firing to avoid sudden expansion which could dislodge the cloisons. Have a long-bladed spatula ready. Withdraw the piece from the muffle as soon as the cloisons start to embed and, before the grounding hardens, press down gently with the sharp end of the spatula blade on the tops of cloisons which are standing too high.
13 From this stage onwards, when cleaning after firing, it is important to check that all fire-scale has been removed from the sides of the cloisons, scouring with a glass brush if necessary.
14 The finely ground, well-washed pastes can now be filled into the design. It is best to use only colours with similar fusing points. The pastes are laid in with a quill or point to give a thin, smooth layer. The side of the metal is tapped to level out the pastes and excess moisture is drawn off with a blotter. The depth of colour is built up in two or three fine layers, separately fired. For the last infill the firing can be a little stronger to give a good lustre to the fused enamel.
15 If a smooth surface is required, the cloisonné work will require filing with carborundum stone, then with emery paper and this is followed by a quick refiring to restore the gloss.

A plaque enamelled in this way is suitable for setting into a frame as a wall decoration. If smoothly finished plaques are made they can be set into a table top or screen.

Two panels from the 'History of Enamelling' series, by Erika Speel.
top *Medieval enamellers* **bottom** *Modern enamellers*

FLANGED PENDANT (OR BRACELET SECTION) WITH PAINTED DECORATION

1 Cut a rectangle 3.5 cm (about 1.5 in) wide by 6 cm (about 2.5 in) long from 20 gauge sheet copper for a pendant, or a proportionately smaller rectangle for a bracelet section.
2 Cut the two top corners of the rectangle away to leave a flange about 1.5 cm (0.75 in) square. For a pendant the plaque can be given an interesting outline. A bracelet section will require a flange at each end.
3 File all the corners to leave them rounded. Anneal the plaque. Bend the flange into a tube with jewellers' round-nosed pliers, so that a channel is left to take a chain or link.
4 Clean and scour the plaque. Brush the reverse side (the side to which the open end of the channel faces) with gum and brush some gum inside the channel, then sieve a hard white enamel over the gum. Dry and fire briefly.
5 When cool and cleaned, brush the top surface with gum and cover with white enamel, dry and fire on a U-shaped support.
6 When cool, stone the top surface to leave it smooth and quite level, and then brush out with a glass brush and water. Apply a second fine layer of white, or ivory, enamel and fire again.
7 Stone smooth and brush out, then refire to restore the glazed surface, without firing too strongly.

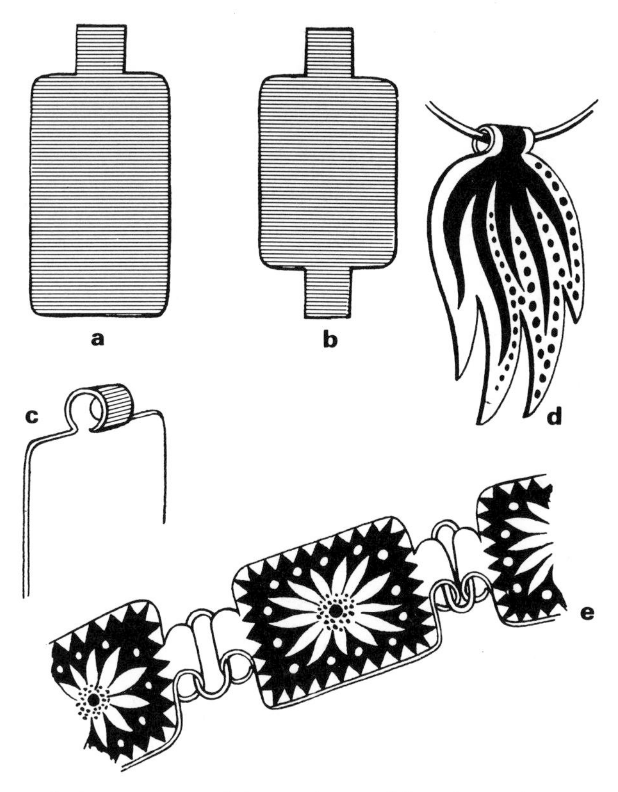

a & b *Small plaques prepared for pendant and bracelet section*
c *The flange is turned over to take a link*
d & e *The plaques after enamelling and painting with overglazes*

8 Prepare the metallic oxides which are to be used for the painted design, working the painting medium and thinner well into the pigments just before use.
9 Apply the painted work by brush. Place the plaque on its support under a lamp or on the top of the hot kiln to dry completely.
10 To fire, first hold the plaque in front of the open muffle until no fumes rise from the surface of the painting and the pigment appears powdery and dry. Then place the work into the muffle, leaving the door ajar for the first few seconds. Watch the firing and remove the work as soon as the pigments have fused into the grounding – the colours will look paler and will be flush with the grounding.
11 When cool, clean the edges of the plaque and apply further pigment to complete or reinforce the painted work, and dry and fire as before.
12 If needed add softer colours and refire with a gentler, shorter firing to fix this layer.
13 When the painted work is complete and no further firing is required the edges of the metal are polished with very fine wet and dry emery paper, filing away from the painted surface.

A SILVER RING WITH TRANSLUCENT ENAMEL

1 Prepare a silver ring, wide enough and of a thick enough gauge to allow a sunken channel to be worked into the outer surface. The ring must have been soldered with enamelling quality solder. (The join can be protected during firing by painting over the inside surface of the ring with a paste made of jeweller's rouge and water, letting this dry before applying enamel.)

2 The recessed channel or groove is engraved or etched into the centre of the ring, leaving a narrow rim at top and bottom. The base of the channel should be left roughened with the graver, to give a better hold to the enamel and to increase the translucent effects of the fired enamel.

3 Prepare transparent turquoise enamel as a finely grained, well-washed paste, adding two drops of gum. Work the paste into the channel, using a point, giving thin cover around the ring. Press on the surface of the enamel with a small pad of linen as each section is filled, to squeeze out air pockets and absorb moisture.

4 Leave the ring standing on a spatula to dry. Pre-heat a flat, mesh firing support in the muffle, remove it from the kiln and carefully place the ring onto the hot support. When the ring has absorbed some heat, re-insert the support with the ring into the muffle. Give a very brief firing, removing the work as soon as the enamel starts to melt and let the ring cool slowly on its support. Reverse the ring for alternate firings from this stage onwards.

5 Apply the second layer of turquoise enamel and refire.

6 The final layer of enamel should bring the paste slightly above the level of the rims. Fire as before.

7 File the enamelled surface smooth and level with the silver rims using carborundum and water. Brush out with a glass brush. Refire to restore the gloss.

8 Clean the silver in pickle and polish the metal surface with care to avoid marking the enamelled area.

A silver ring enamelled with translucent turquoise over a simple engraved pattern.

5 Some Common Problems

BLISTERS

Large blisters may form if an unsuitable metal base (e.g. an alloy) has been used or where solder underlies the enamel. Large, sagging blisters may form when thickly applied counter-enamel pulls downwards in a slow firing. A fine bubble effect in the surface is due to enamel powder having been too loosely applied or when the powder has been sieved from too great a height. Large blisters should be broken open with a metal point, smoothed by filing and rinsed, then refilled with enamel before refiring. Small bubbles tend to disappear with refiring.

CRACKS

When there is extensive cracking parts of the glaze may spring off the metal base. Cracks can be avoided by applying the correct thickness of enamel and by giving a counter-enamel where there is tension between the metal and enamel. Acute angles and straight-sided plaques of thin metal should be avoided to reduce stress which leads to the formation of cracks. Cracks due to too rapid cooling can be repaired by refiring and slow cooling.

CRAZING

Crazing or hairlines will appear if the grounding expands during refiring and pulls the superimposed pigment apart. This is particularly evident on larger scaled work painted with metallic oxides. Choosing the correct firing sequence and warming the work before refiring reduces this tendency.

DULLNESS OF COLOUR

Some enamels improve in brilliance after two firings. Some colours tend to lose their tone if fired too often or at too high a temperature. Dullness in transparents may be due to underfiring, which can be rectified by a brief refiring at a higher temperature. Dullness may be due to insufficient washing of the paste or lack of brightness of the metal base when transparents have been applied. Not all enamel colours are equally brilliant and some become dulled if immersed in pickle or acid. Dull patches may be due to insufficient brushing out and rinsing after stoning of the fired surface.

EDGES

Enamel tends to shrink a little during firing. On smoothly surfaced metal and shaped articles the enamel may shrink away from the rims of the work, leaving untidy or darkened edges. Overfiring aggravates this problem. Good cover should always be given at the edges of the metal surface, particularly if a smooth metal base is used. Where possible shaped work should be fired in a reversed position in alternate stages to prevent flow to or away from the edges of the article.

FIRE-SCALE

When copper and alloyed bases are used the bare metal areas become covered with an oxidized layer due to firing. This deposit is known as fire-scale. The scale breaks up into tiny fragments and if any become embedded into the molten enamel a dull or black spot will be left. Fire-scale should be removed from the metal after each firing to avoid this. Scale-inhibiting coatings can be painted onto the bare metal areas but the edges of the work will still require cleaning for every firing. If lightly embedded into the enamel surface, a speck of fire-scale can be scraped off with an old metal graver or sharp steel point. If deeply embedded the blemish has to be stoned off with carborundum.

UNEVEN ENAMEL COLOUR

Some colours are very sensitive to small changes in temperature and may emerge from the firing with a degree of shading due to the lesser heat in the muffle at the door-end. To give equal firing the work has to be given a half turn during the firing.

UNEVEN GLAZE

If the fired surface of the enamel is not smooth and shiny all over it may be due to underfiring, which shows as a gritty or orange peel surface, or the disparity may be due to overfiring where the work is too near the heat source at one end while the near-side is still cooler. Partial overfiring may be difficult to rectify and is generally due to firing too large an article for the size of the muffle. If the glaze emerges from the firing having drawn from the metal base in patches this may be due to overfiring or to the use of a greasy or incompatible metal base. Scouring the metal surface to roughen it a little can help to prevent the drawing together and this treatment is particularly beneficial with silver and silver foil bases.

WARPING

Warping of the metal base is due to the pull of the fired enamel on the metal which causes tension as the materials cool at different rates. The thicker the covering of enamel is, the greater the pull on the metal. The stress is reduced by applying a counter-enamel to the reverse side of the work or by using a heavier gauge of metal and a thinner covering of enamel. The distribution of enamel over the surface of the work should be as even as possible. Flat plaques should be weighted down during cooling to prevent warping. Where possible the metal base should be shaped to increase rigidity. Bowls retain their shape more successfully if the sides are gently curved.

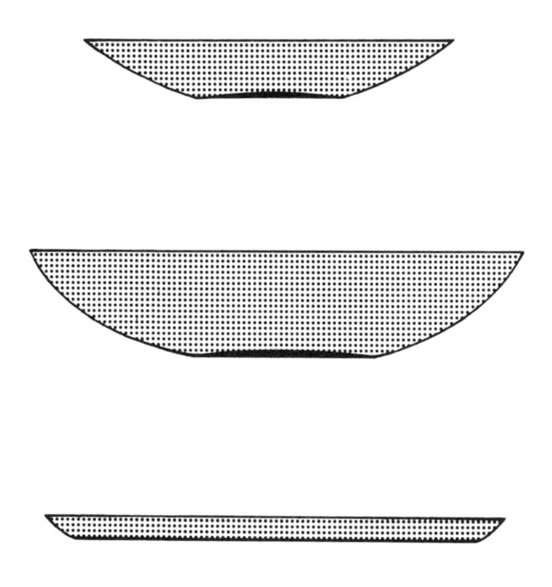

Raised dishes should have gently curved sides if possible

Appendices

Firing Temperatures

The following are approximate figures, as results vary according to length of firing time and differences in materials.

Soft-fusing enamels	750–790°C	1382–1450°F	(muffle glows cherry red)
Medium-fusing enamels	790–820°C	1450–1508°F	(muffle glows bright cherry red)
Hard-fusing enamels	820–850°C	1508–1562°F	(muffle glows bright red)
Higher setting for large or heavy gauge metal bases	950°C	1742°F	(muffle glows orange red)
Recommended firing temperatures on steel	800–820°C	1472–1508°F	
Transfers fired to a grounding	650–700°C	1202–1292°F	
Liquid precious metals	700–730°C	1292–1346°F	
Metallic oxide overglazes	720–750°C	1328–1382°F	

Glossary

Alloy: A metal compound produced by melting two or more metals together.
Annealing: Removing hardness and brittleness by heating to dull red and then cooling.
Basse taille: Translucent enamel fired over a metal surface which has been recessed to varying depths.
Bronze: An alloy of copper with tin.
Burnishing: Brightening and smoothing gold or silver with a very smooth metal or agate tool.
Calx: The mixture added to transparent enamel to make it opaque (generally calcium with lead and tin).
Carborundum: Silicon carbide – a very hard abrasive.
Champlevé: Sunken cell enamelling: the cells are worked into the surface of the metal to an even depth with metal divisions left between the cells to form the outlines of the design. Enamels are fired into the recessed cells, then polished smooth.
Chasing: A goldsmithing technique of decorating the surface of the metal or modelling it by the use of special punches.
Cloisonné: Raised cell enamelling: compartments or enclosed areas are made by attaching narrow, shaped strips of metal or wires to a metal base, then filling the cells formed in this way with enamels.
Cloisons: The fine metal divisions or outlines in the cloisonné technique.
Copper: A pinky-brown metal. Melting point 1083°C (1981.4°F).
Counter-enamel: The layer or layers of enamel fired to the reverse side of a metal article which is decorated with enamel on the top surface, to equalize the stress on the metal.
Cracks: Fractures or splits in the enamel surface.
Draw plate: A steel plate with graduated holes through which annealed wire is drawn to change its diameter.
Dusting (sieving, sifting): Applying finely pulverized dry enamel powder through a screen, mesh or sieve (sifter) to deposit a thin layer or coating of enamel.
Enamel: An easily fusible glass which will bond to certain metals if fused at the correct temperatures, consisting of a glassy base coloured by the addition of metallic oxides and salts.

Encrusted enamel: Enamel fired to the surface of metal.
Engine turning: Giving a sharp, bright cut to metal surfaces by means of special lathes.
Engraving: Making recessed designs in the metal surface by cutting away or carving out parts of the metal to a controlled depth.
Etching: Removing parts of the metal surface to a controlled depth by means of acids.
False cloisonné: Embedding a complete wire or metal design into a grounding of enamel to simulate cloisonné.
Filigree: Similar to cloisonné enamelling but with simple designs formed from round, beaded, or twisted wires.
Firing: Bringing the enamel-covered metal to sufficient heat to melt the enamel and let it fuse to the metal base.
Fire-scale: The oxidized layer which forms on copper and its alloys as a result of heating.
Flux: The glassy, uncoloured base of enamel, also known as fondant. (Flux used in connection with soldering of metals refers to the medium which helps the solder to flow, i.e. borax.)
Fluxing (crowning): Adding one or two very fine layers of flux over coloured enamels or overglazes.
Frit: Lumps or rough chunks of enamel.
Gilding: Covering with a fine layer of gold or silver.
Gilding metal: Copper alloyed with zinc. If proportion of zinc less than 10%, melting point is 990°C (1814°F).
Gold: Pure 24 carat gold is yellow in colour and it is very soft, with a melting point of 1063°C (1945.4°F). The melting point of 18 carat gold (yellow) is 905°C (1661°F).
Grinding: Pulverizing lump enamel.
Grisaille: Monochrome colour scheme, usually black and white enamel, with half-tones produced by varying the thickness of the superimposed layer.
Gum: A mucilage or holding agent.
Hard (high-fusing) enamel: Enamel which requires at least 800°C (1472°F) to fuse.
Key: The roughening or engraving of a surface.
Kiln: A furnace with an inner chamber which can be heated to and retain the temperatures needed for the firing of enamelwork.
Limoges-style painted enamels: Painted enamels created with coloured pastes applied over a prefired enamel grounding, the compositions being built up in stages, working from the hard enamels to the softer fusing ones.
Liquid metals: Metals which have been pulverized and mixed with a medium so that the powder is held in suspension.
Metallic oxides: Oxides of metals which are used as colouring agents in the making of enamels or used pure as painting pigments or overglazes.
Miniature painted enamels: Small painted works composed of an enamel grounding overpainted with layers of metallic oxide overglazes.

Muffle: The inner chamber of a kiln or furnace in which enamelwork is placed during the firing.
Opalescent enamel: A specially formulated type of enamel which can produce an irridescent or milky sheen when fired.
Opaque enamel: Enamel which is dense and through which light cannot pass so that the underlying metal base or colours cannot show through.
Overglazes: Very finely powdered metallic oxides or enamels applied and fired over an enamel grounding.
Oxidization: The discolouration which appears on base metals when exposed to oxygen in the air.
Paillons: Shaped pieces of gold or silver foil.
Painted enamels: Enamelwork produced by applying finely powdered enamels or overglazes over a grounding of white or other opaque enamel.
Pickle: Acid diluted in water.
Planish: A silversmithing technique of smoothing the metal surface with a special hammer.
Plaque: A pre-shaped, prepared piece of metal.
Plique-à-jour: A technique of enamelling in which translucent colours are fired to remain suspended between outlines or ribs of metal, without a metal backing under the enamel, allowing light to pass through.
Pyrometer: A gauge which can register the temperature inside the muffle.
Repoussé: A goldsmithing technique of working a relief design into the surface of metal with a hammer and special punches.
Sand cushion: A flat, circular, leather pad filled with sand used as a resilient surface for metal working.
Scrolling: A method of making a free-form design with enamel while it is in a molten state.
Sgraffito: Scratching through an applied layer of enamel before it has been fired, to expose parts of the grounding.
Silver: A white-coloured, very reflective metal, which is soft and malleable. The melting point of fine silver is 961.5°C (1762.7°F) and sterling silver melts at 893°C (1639°F).
Soldering: Joining pieces of metal by fusing them together with an alloy which has a lower melting point.
Stoning: Filing with an abrasive material, such as a carborundum stone, to smooth the surface of fired enamel.
Supports: Holders, trivets, stilts, planches or cradles of metal or clay on which enamelwork is placed for firing.
Tragacanth: A plant from which an extract can be prepared which makes a suitable enamelling gum.
Transfers: Metallic oxide lithographs made commercially which can be applied and fired to a prefired enamel ground.
Translucid enamel: A term used to describe basse taille enamelling or transparent or translucent enamels.
Transparent enamel: Enamels which allow light to pass through.
Tripoli: A fine abrasive powder used as a polishing compound.

Bibliography

BATES, Kenneth, *Enamelling*, World Publishing Co., Cleveland and New York, 1951

BROWN, W.N., *The Art of Enamelling on Metal*, Scott, Greenwood & Son, London, 1900

CLARKE, Geoffrey and FEHER, Francis and Ida, *The Technique of Enamelling*, B.T. Batsford Ltd, London, 1967

CRAWFORD, John, *Introducing Jewelry Making*, B.T. Batsford Ltd, London, 1969

CUNYNGHAME, H., *Art Enamelling on Metal*, A. Constable & Co., London, 1899

DAWSON, Mrs Nelson, *Enamels*, Methuen & Co. Ltd, London, 1906

DAY, Lewis F., *Enamelling*, B.T. Batsford Ltd, London, 1907

FAIRFIELD, Del, *Enamelling*, Hodder & Stoughton, London, 1977

FISHER, A., *Enamelling on Metal*, The Studio, London, 1906

GEE, George E., *The Goldsmith's Handbook*, Crosby Lockwood & Son, London, 1918

GODDEN, Robert and POPHAM, Philip, *Silversmithing*, Oxford University Press, London, 1971

HARPER, William, *Step by Step Enamelling*, Golden Press, New York, 1973

KRONQUIST, Emil F., *Metalwork for Craftsmen*, Dover Publications, New York, Reprint 1972

MARYON, Herbert, *Metalwork and Enamelling*, Dover Publications, New York, Reprint 1971

MILLENET, Louis-Elie, *Enamelling on Metal*, trans. H. de Koningh, The Technical Press Ltd, Kingston Hill, Surrey, 1951

NEVILLE, Kenneth, *The Craft of Enamelling*, Mills & Boon, London, 1966

PACK, Greta, *Jewelry and Enameling*, D. van Nostrand Co., New York, Reprint 1941

ROTHENBERG, Polly, *Metal Enamelling*, Allen & Unwin, London, 1969

SEELER, Margaret, *The Art of Enameling*, Galahad Books, New York, 1969

SWINKLES, Bep, *Enamelling*, Robert Hale, London, 1970

UNTRACHT, Oppi, *Enameling on Metals*, Greenberg, New York, 1957

WILSON, Henry, *Silverwork and Jewellery*, Pitman Publishing, London, Reprint 1973

WINTER, E., *Enamel Art on Metals*, Watson Guptil, New York, 1958

Suppliers

The following firms supply enamels, painting enamels (metallic oxide overglazes), enamelling accessories. They also offer extensive ranges of ancillary materials, tools and equipment such as sheet metal and/or shaped blanks, wire, solders, etching materials, transfers, liquid precious metals, foils, metal-working/jewellery-making/enamelling tools and equipment, kilns and firing equipment, and cleaning materials. Firms marked ** can supply catalogues.

UK

W.G. Ball Ltd**
Anchor Road, Longton, Stoke-on-Trent, England ST3 1JW
Tel: Stoke-on-Trent (0782) 373956/312286

Range of lead-free enamels and enamels for steel

Craft O'Hans (London)**
21 Macklin Street, London WC28 5NH, England
Tel: 01 242 7053

Range includes presentation articles ready for enamelling

Craft O'Hans (Wales)**
The Old Mill, Denbigh, Nannerch, near Mold, Clwyd, North Wales
Tel: 03528 542

Deancraft Ltd**
Lovatt Street, Stoke-on-Trent ST4 7RL, England
Tel: 0782 411049/46175/410117

Pat Johnson Enamels**
51 Webb's Road, London SW11, England
Tel: 01 228 0011

UK agent for American enamels and accessories

C.B. Latham (Colours) Ltd**
Clifton Street, Fenton, Stoke-on-Trent ST4 3QQ,
England
Tel: 0782 46352

UK agent for Austrian enamels

Peter H. Wolfe & Gudde Jane Skyrme Workshops
84 Camden Mews, London NW1, England
Tel: 01 267 4979

Selection of enamelling and jewellery materials.

USA

Allcraft Tool & Supply Co.**
P.O. Box 723, 100 Frank Road, Hicksville, New York 11801, USA
Tel: 516 433 1660

Range is extensive but does not include overglazes or liquid metals

The Ceramic Coating Co.**
Head Office: P.O. Box 370 Newport, Kentucky 41072, USA
Tel: (606) 781-1915
Regional stockists in Arizona, California, Connecticut, Florida, Hawaii, Illinois, Kansas, Louisana, Maryland, Massachusetts, Michigan, Minnesota, Missouri, New Mexico, New York, North Dakota, Ohio, Oregon, Pennsylvania, Texas, Virginia, Washington and Wisconsin.

Thomas C. Thompson enamels and enamelling accessories

Minthorne International Co. Inc.**
55 North Main Street, P.O. Box 1000, Freeport, New York 11520, USA
Tel: (516) 546-4700

Thomas C. Thompson enamels and enamelling accessories

Quimby & Co.
60 Oakdale Road, Chester, New Jersey 07930, USA

Metal-cleaning solutions included in the range

Sherry's Western Ceramic Supply Co.**
948 Washington Street, San Carlos, California 94070, USA

A comprehensive range (etching materials not stocked)

Suppliers of special materials and equipment, including craft tools:

UK

Fred Aldous Ltd
P.O. Box 135, 37 Lever Street, Manchester M60, England
Tel: 061 236 2477

Range includes kilns, copper blanks, painting enamels

Dryad**
P.O. Box 38, Northgate, Leicester, England
Tel: 0533 50405

Range includes kilns and artists' tools, some enamels

Flamefast Ltd
Pendlebury Industrial Estate, Manchester, England
Tel: 061 793 9333

Kilns and gas equipment for enamelling

Technical Leisure Centre**
1 Grangeway, London NW6 2BW, England
Tel: 01 328 3128

Propane gas torches and burners, and craft tools

J. Gordon Parks & Partners**
Head Office (London Bullion Ltd)
73 Farringdon Road, London EC1, England
Tel: 01 405 4441/01 242 0155
Shop: 193 Wardour Street, London W1V 4LP, England
Tel: 01 439 2349

Comprehensive range of tools and equipment for silversmithing and jewellery making, etching, cleaning and polishing materials and equipment, kilns and accessories.

Copper and other non-precious non-ferrous metals:

J. Smith & Sons (Clerkenwell) Ltd**
Head Office: 42–44 St John's Square, London EC1P 1ER, England
Tel: 01 253 5937

Sheet metal and shaped blanks

Branches: Biggleswade, Bedfordshire; Birmingham; Bracknell, Berkshire; Bristol; Chelmsford, Essex; Cleckheaton, Yorks; Gateshead, Tyne-Wear; Horsham, Sussex; Maidstone, Kent; Manchester; and Nottingham.

Precious metals:

Blundell & Sons Ltd**
199 Wardour Street, London W1V 4JN, England
Tel: 01 437 4746

Range includes silver, silver jewellery findings and enamelling solder

Johnson Matthey Metals**
Head Office: 100 High Street, Southgate, London N14 6ET, England
Tel: 01 882 6111
Retail supplies: 43 Hatton Gardens, London EC1, England

Gold and silver and liquid precious metals

George M. Whiley Ltd
The Runway, Station Approach, South Ruislip, Middlesex, England
Tel: 01 841 4241

Precious metal foils

USA

Most of the firms listed as suppliers of enamels and enamelling requisites also stock a wide range of accessories.

T.B. Hagstoz & Sons** 709 Sansom Street, Philadelphia, Pennsylvania 19106, USA	*Precious metal foils and liquid precious metals. All gold alloys, silver and solders. Jewellery findings, metal working tools, etching materials. Shaped blanks in copper and silver and pre-enamelled steel.*
Minthorne International Co. Inc.** 55 North Main Street, P.O. Box 1000, Freeport, New York 11520, USA Tel: 516 546 4700	*Precious metal foils and liquid. Precious metals. Jewellery findings*

BUSINESS ADVICE

UK

A pamphlet *Starting Your Own Business* is available from branches of Lloyds Bank, or from Lloyds Bank Plc, 71 Lombard Street, London EC3 3BS.

USA

Two pamphlets *Steps to Starting a Business* and *The Handcrafts Business* are available from branches of The Bank of America, or from Small Business Reporter, Bank of America, Department 3120, P.O. Box 37000, San Francisco, CA 94137.

GUILDS AND NEWSLETTERS FOR ENAMELLERS

UK

The Guild of Craft Enamellers publishes regular newsletters. Guild information and regional addresses from Richard Fox, Renarden, Lyth Hill, Shrewsbury, Salop, or Maureen Carswell, Sainthill, 33 Kingsland Road, Shrewsbury SY3 7LB.

USA

Glass on Metal (enamelists' newsletter), published six times every year. Details from the Ceramic Coating Co., P.O. Box 370 Newport, Kentucky 41072. This newsletter contains details of regional guilds.

Index